Rocky Landscapes of the World with Google Earth

Andrew Goudie

Rocky Landscapes of the World with Google Earth

Springer

Andrew Goudie
Portland, UK

ISBN 978-3-031-77417-1 ISBN 978-3-031-77418-8 (eBook)
https://doi.org/10.1007/978-3-031-77418-8

This Springer imprint is published by the registered company Springer Nature Switzerland AG
The registered company address is: Gewerbestrasse 11, 6330 Cham, Switzerland

This book is dedicated to my two grandchildren,
Teddy and Esme

Acknowledgements

Over the years I have been able to study rocky landscapes in many parts of the world. Among those I would like to thank are: Tex Reeves and Gordon Wells in the USA High Plains; Celia Nyamweru in Kenya; Heather Viles in the Colorado Plateau, Chile, China, Morocco, Jordan and Namibia; Members of the Kimberley Research Project in Australia; Members of the Bahrain Surface Materials Resources Survey in Bahrain; Members of the Wahiba Sands Project in Oman; Members of the International Karakoram Project in Pakistan; Ken Coombe and Denys Brunsden in Dorset; Max Gibbons, Andrew Watson and David Price Williams in Eswatini; John Wellington in the Transvaal; Andrew Watson and Ken Pye in Kenya; Stephen Stokes and Larry Barham in Zambia; David Thomas, Ken Pye and Andrew Watson in Zimbabwe; Nick Rosser, Bob Allison and Piotr Migoń in Jordan; Dick Grove in Africa; Bridget and Raymond Allchin, K. T. M. Hegde, Charles Sperling, R. S. Pappu and S. N. Rajaguru in India; Mike Meadows, Rodney Maud, Tim Partridge, and Stefan Grab in South Africa; Marjorie Sweeting in the UK and Ireland; Harald Schnepfleitner, Stefanie Fruhmann and Oliver Sass in Morocco; Asma Al-Farraj in the UAE; Nina Piggott in Shrophire; Rita Gardner in various parts of England and Wales and South India; George and Terence Adamson in Kora; Stephen Stokes, Mahmoud Ashour and Nabil Embabi in Egypt; Qinglin Guo, He Qing Huang, Wenwu Chen and Hong Zhang in China; Piotr Migoń and Filip Duszyński in Poland and the Weald; Mike Wood in Cyrenaica; Jan Kalvoda in Czechia; Mick Day in the Kimberley, Oxfordshire and Death Valley; Adrian Parker in Oxfordshire, Oman, Eswatini and the UAE; Ash Parton, Gareth Preston and Derek Kennett in Oman; Allan Pentecost and Karl-Heinz Pfeffer in Swabia; Andrew Watson, Jacques Bonvallot and Jacques Viellefon in Tunisia; Jean Grove in the Valais of Switzerland; V. Ranov in Tajikistan; Dan Yaalon and Asher Shick in Israel; Lew Watts in Botswana; and Mike Summerfield in South Africa, Botswana and England. At Cambridge, I benefitted greatly from the lectures of Bruce Sparks on rocks and relief. Gordon Wells and Paolo Paron, on visits to Oxford, introduced me to the value of satellite images in geomorphological research, while Marjorie Sweeting attracted many international karst scholars to Oxford, including a group from China led by Yuan Daoxian, whom it was my pleasure to meet and listen to.

It would have been impossible to write this book, which gave me relief from dealing with the incomparably bad customer services of British Gas, without the free and incomparably good resources of Google Earth and Google Scholar. I am also grateful to the late Marjorie Sweeting, Heather Viles and Alice Goudie for some of the images. I also owe a debt to Dr. Robert Doe and Mr. Deivasigamani Arumugam of Springer. The former encouraged me to write this book, and the latter carefully saw it through to production.

Contents

About the Author

Portrait of the author at St. Cross College by Dr. Kate Scott

Andrew Goudie, Ph.D., D.Sc. Emeritus Professor of Geography at the University of Oxford, former Pro-Vice-Chancellor, former Director of the China Centre, Honorary Fellow of Hertford College and the former Master of St Cross College, is a geomorphologist. He is a recipient of a Royal Medal from the Royal Geographical Society, the Mungo Park Medal of the Royal Scottish Geographical Society, and the Farouk El-Baz Award of the Geological Society of America. He is a Fellow of the British Society of Geomorphologists. He has been Chair of the British Geomorphological Research Group, President of the Geographical Association, and President of the International Association of Geomorphologists. He has written extensively on geomorphology in some 50 books and 300 scientific papers. He is the author of among others *The Landforms of England and Wales* (1990), *Great Warm Deserts of the World* (2002), *Arid and Semi-arid Geomorphology* (2012), *Great Desert Explorers* (2016), *The Human Impact* (2018), *Landscapes and Landforms of England and Wales* (with Piotr Migoń) (2020), *Camel Tracks* (2021), and (with Denys Brunsden), *Geological Pioneers of the Jurassic Coast* (2023). Recently he has used Google Earth in two companion volumes: *Desert Landscapes of the World with Google Earth* (2022) and *Landscapes of the Anthropocene with Google Earth* (2023).

Introduction

1

Abstract

This chapter starts with an introduction to Google Earth and assesses its value for investigating landscapes since it was launched in 2005. This is followed by a discussion of previous work on the relationships between rocks and landscapes and a listing of relevant UNESCO World Heritage Sites, American National Parks and geosites.

Keywords

Google Earth · Rock · Geomorphology · Landscape · Geoasthetics

1.1 Google Earth

The rationale of this book is to provide carefully selected images from Google Earth, to discuss what they show, and to provide references to the sites portrayed so that readers can go deeper if they wish. Google Earth (GE) is remarkably useful for examining and appreciating phenomena on the face of the Earth. This book aims to demonstrate its value for understanding the landscapes associated with particular rock types and structures and for appreciating their aesthetics. Many GE images are extraordinarily beautiful and this volume is, in a sense, a contribution to Geoasthetics. I hope that any students who browse through its images may be stimulated to pursue Geomorphology. The book does not aim to be a comprehensive treatise on the links between rocks and relief (for which see Migoń 2025).

GE was launched in 2005 and maps the Earth by superimposing satellite images, air photographs and GIS data onto a 3D globe, allowing its users to see landscapes from a range of angles (Liang et al. 2018). It is regularly updated and in its basic form is free for users. It is easily accessed via personal computers. Imagery resolution ranges from 15 cm to 15 m. It provides a distance measuring tool and it is also possible to obtain elevation cross profiles. GE provides near global coverage, so that it is immensely important for determining global distributions of particular phenomena (Goudie 2020; Traganos et al. 2018), including miscellaneous desert landforms (Goudie 2022a) and Anthropocene features (Goudie 2023b). However, in general, the resolution is not adequate for the identification of many micro-forms, so that most of the features illustrated in this book are fairly substantial in size. There has also been a tendency to select many images from drylands, as it is in these environments that structures and forms are less obscured by vegetation. This means, for instance, that some of the wonderful sandstone landscapes in southern Poland and neighbouring parts of Czechia, which are for the most part forested, are not tiven the attention here that they deserve.

GE (https://www.google.co.uk/intl/en_uk/earth/) (Accessed 7th June 2023) has been much used in landscape studies, for both teaching (see Lisle 2006; Dolliver 2012; Fisher et al. 2012; Demirci et al. 2013; Tooth 2015; Ortiz-Guerrero and Loizzo 2024) and research (Scheffers et al. 2015). Table 1.1 lists selected studies of the use of GE in the study of particular geomorphological phenomena. A useful site to see GE images of landslides is https://blogs.agu.org/landslideblog/ (Accessed 2nd May 2023).

In this volume, the latitudes and longitudes of images are provided, so that they can be re-located by the reader, as are scale bars. These are located in the bottom right corner of all images. Only vertical images are used in this book, but it is also possible to tilt them.

1.2 Rocks and Geomorphology

The terrestrial Earth surface is covered by 64% sedimentary rocks (a third of which are carbonates), 13% metamorphics, 7% plutonics, and 6% volcanics. Significant regional

A. Goudie, *Rocky Landscapes of the World with Google Earth*, https://doi.org/10.1007/978-3-031-77418-8_1

Table 1.1 Examples of the use of Google Earth in geomorphological studies

Channel and drainage patterns (Lee 2015; Colombera and Mountney 2019; Kotluri et al. 2024)
Coasts (Scheffers et al. 2012; Warnasuriya et al. 2020)
Conglomerate landscapes (Migoń 2020)
Debris flows (Xu et al. 2011; Dias et al. 2022)
Denudation (Cendrero et al. 2022)
Dunes (Goudie 2020, 2023a, c; Goudie et al. 2021a, b; Goudie 2022a, b)
Dyke swarms (Samal et al. 2019)
Estuaries (Ezcurra et al. 2019; Baum et al. 2024)
Faults, folds and fractures (Tewksbury et al. 2012; Lageson et al. 2012)
Glacier retreat (Nainwal et al. 2016)
Greenstone terrain and dyke swarms (Anhaeusser 2024)
Gully erosion (Karydas and Panagos 2020; Reece et al. 2023; Shirani et al. 2023; Ma et al. 2024; Liu et al. 2024)
Lakes (Scheffers and Kelletat 2016)
Landslides (Fisher et al. 2012; Rabby and Li 2019; Peng et al. 2021; Sati et al. 2022; Sun et al. 2023; Bhandari et al. 2024; Ali 2024)
Lineaments (Jonathan et al. 2024)
Mangrove swamps (Singh et al. 2024)
Mud volcanoes (Komatsu and Feyzullayev 2024)
Permafrost landforms (Whalley 2021)
River morphology and junctions (Henshaw et al. 2020; Pereira et al. 2024)
Rockfalls (Swanger and Admassu 2018)
Rock Glaciers (Schmid et al. 2015; Charbonneau and Smith 2018; Pandey 2019; Harrison et al. 2024)
Salt marshes (Goudie 2013; Sarmiento et al. 2020)
Soil erosion (Boardman 2016)
Thaw slumps (Luo et al. 2019)
Tiger bush (Brousse tigrée) (Paron and Goudie 2007)
Volcanic landforms (Gabrielli et al. 2020)
Yardangs (Yang et al. 2023)
Wetlands (Amani et al. 2019)

differences in their distribution are observable: e.g., carbonate-containing rock units represent 40% of the area in Europe while they represent only 14% in South America (Hartmann and Moosdorf 2012).

The importance of rock as a control of landform development has long been appreciated but seldom quantified. It was included as 'structure' in the Davisian trilogy of 'structure, process and stage' (Davis 1899), and many geomorphologists have seen landforms as the result of the interaction of force (processes) and resistance (material properties). An excellent treatment of rocks and associated landscapes from the Davisian era is the superbly illustrated textbook by Lobeck (1939) wqith its wonderful block diagrams of folds, faults and other structures.

However, when the quantitative revolution in geomorphology arose in the 1960s, the study of processes rather than resistance came to dominate much Anglo-Saxon geomorphology, and a plethora of reductionist studies of processes was undertaken. Nonetheless, some of the prominent practitioners of this revolution were aware of the importance of rock control and, for example, Schumm (1956) was concerned with controls on badland slope form, while Chorley (1959, 1964) used a penetrometer in an attempt to assess the resistance of an escarpment-forming rock in southern England. Nevertheless, the number of studies that have attempted to quantify rock properties and to relate these to landforms has been relatively small. There have been some important but little heeded calls for quantitative studies of material controls in geomorphology, most notably by Yatsu (1966) and Whalley (1976), while, atypically and presciently, Selby (1993) pioneered the use of geotechnical studies in his analysis of slope forms. Recently, there has been a growth in quantitative techniques for analysis the role of rocks in landscape development, and these are reviewed in Goudie (2016). There has, for example, been a considerable growth of interest in the links between river channel morphology and rock resistance (e.g. Bursztyn et al. 2015; Chen and Byun 2023). In a similar vein, Duszyński et al. (2021) recount the study of rock properties and rock-controlled landforms during the second half of the twentieth century, following on from earlier reviews by Gerrard (2008) and Bishop (2008).

There have been some essentially regional studies that have explored the influence of rocks of different ages and types on scenery. In England mention can be made of Stamp (1946), Miller (1953) and Whittow (1992). There have also been many essentially qualitative studies of the geomorphology of particular rock types, including those of Sparks (1971) and Gerrard (1988), and detailed texts on limestone relief (e.g. Sweeting, 1972; Jennings, 1985), sandstones (Young et al. 2009), and granite (e.g. Twidale and Vidal Romani 2005; Migoń 2006). Rock types and structures are also a primary control of the nature and occurrence of many mass movements, including landslides (Brideau and Roberts 2022). There are also several works devoted to rocky coastlines (e.g. Trenhaille 1987; Sunamura 1992). Finally, structures associated with tectonic forces, and their associated landforms, are reviewed by Keller and Pinter (2002) and Burbank and Anderson (2001). These are discussed in the next chapter. The remarkable World Geomorphological Landscapes series (edited by Migoń and published by Springer) now has a large number of regional studies which include discussions of the impacts of rocks on relief.

Table 1.2 Examples of World Heritage Site rocky landscape types discussed in this book

Aeolian Islands (Italy)
Air and Ténéré (Niger)
Belize Barrier Reef (Belize)
Canaima (Venezuela)
China Danxia (China)
Cliff of Bandiagara (Mali)
Dorset and East Devon Coast (UK)
Ennedi Massif (Chad)
Göreme National Park and the Rock Sites of Cappadocia (Türkiye)
Great Barrier Reef (Australia)
Ha Long Bay (Vietnam)
Hawaii Volcanoes (USA)
Hierapolis-Pamukkale (Türkiye)
Huanglong (China)
Iguazu/Iguaçu (Brazil and Argentina)
Maloti-Drakensberg (Lesotho and South Africa)
Meteora (Greece)
Mosi-oa-Tunya—Victoria Falls (Zambia and Zimbabwe)
Mount Etna (Italy)
Mount Pelée and pitons (Martinique)
Nahanni National Park, Canada
Ngorongoro (Tanzania)
Pitons, cirques and ramparts (Réunion)
Plitvice (Croatia)
Purnululu (Australia)
Rock Islands Southern Lagoon (Palau)
Serengeti National Park (Tanzania)
Simien National Park (Ethiopia)
South China Karst (China)
Tassili n'Ajjer (Algeria)
Teide National Park (Spain)
Tongariro (New Zealand)
Tsingy de Bemaraha (Madagascar)
Uluru-Kata Tjuta (Australia)
Volcanoes of Kamchatka (Russia)
Wadi Rum (Jordan)
Western Ghats (India)

1.3 Rocky Landscapes as World Heritage Sites and Geosites

Many of the landscape examples given in this book have been made into World Heritage Sites by UNESCO (https://whc.unesco.org/en/list/?&type=natural) (Accessed 10th June 2024). This is because of their superlative nature both scientifically and aesthetically. Examples are given in Table 1.2.

Many other sites are designated as Geosites, Geoparks, Geomorphosites, or as National Parks, including those in the USA—Canyonlands, Crater Lake (Oregon), Joshua Tree, Zion Canyon, Mesa Verde, Bryce Canyon, Death Valley, Mount St Helens, and Arches. A well-illustrated guide to Geosites that provides insights into a number of locations in this book, including the Colorado Plateau, Monument Valley, Arches, Ngorongoro, Oldoinyo Lengai, Etna, Stromboli, Santorini, Meteora, and Pamukkale, is that by Scoon (2021). UNESCO Global Geoparks mentioned in the present volume include Ngorongoro, Bohol, the Burren, Toba, Zhangye, and Zhangjiajie.

References

Ali A (2024) Landslide susceptibility mapping using modified frequency ratio method in Correb area, South Wollo, North-Western Ethiopia. Discover Geosci 2(1). https://doi.org/10.1007/s44288-024-00053-x

Amani M et al (2019) Canadian wetland inventory using Google Earth engine: the first map and preliminary results. Remote Sens 11(7). https://doi.org/10.3390/rs11070842

Anhaeusser CR (2024) Satellite imagery reveals the nature of the Archaean granite-greenstone basement of the Kruger National Park, South Africa, including the newly discovered Olifants River Greenstone Belt. S Afr J Geol 127(2):163–180

Baum MP, Kennedy DM, McSweeney SL (2024) Large structural estuaries: their global distribution and morphology. Geomorphology 461:109309

Bhandari S, Ghimire M, Mahat P (2024) Landslide susceptibility mapping and vulnerability assessment in the Badigad Watershed of Western Nepal. J Inst Sci Technol 29(1):59–73

Bishop P (2008) Tectonic and related landforms. In: Burt TP, Chorley RJ, Brunsden D, Cox NJ, Goudie AS (eds) The history of the study of landforms, vol 4. Geological Society, London, pp 55–105

Boardman J (2016) The value of Google Earth™ for erosion mapping. CATENA 143:123–127

Brideau MA, Roberts NJ (2022) Landslides in bedrock. In: Shroder JF, Davies T, Rosser N (eds) Landslide hazards, risks, and disasters. Elsevier, Amsterdam, pp 43–97

Burbank DW, Anderson RS (2001) Tectonic geomorphology. Blackwell Science, Malden, Mass., pp 274

Bursztyn N, Pederson JL, Tressler C, Mackley RD, Mitchell KJ (2015) Rock strength along a fluvial transect of the Colorado

Plateau–quantifying a fundamental control on geomorphology. Earth Planet Sci Lett 429:90–100
Cendrero A et al (2022) Denudation and geomorphic change in the anthropocene; a global overview. Earth Sci Rev 233:104186
Charbonneau AA, Smith DJ (2018) An inventory of rock glaciers in the central British Columbia Coast Mountains, Canada, from high resolution Google Earth imagery. Arc Antarc Alp Res 50(1):e1489026
Chen H, Byun J (2023) Effects of erosional resistance on bedrock channel occurrence and morphology: examination of the Seo River catchment in South Korea. Geomorphology 438:108810
Chorley RJ (1959) The geomorphic significance of some Oxford soils. Am J Sci 257:503–515
Chorley RJ (1964) Geomorphological evaluation of factors controlling shearing resistance of surface soils in sandstone. J Geophys Res 69:1507–1516
Colombera L, Mountney NP (2019) The lithofacies organization of fluvial channel deposits: a meta-analysis of modern rivers. Sed Geol 383:16–40
Davis WM (1899) The geographical cycle. Geogr J 14:481–504
Demirci A, Karaburun A, Kılar H (2013) Using Google Earth as an educational tool in secondary school geography lessons. Int Res Geogr Environ Ed 22(4):277–290
Dias VC, McDougall S, Vieira BC (2022) Geomorphic analyses of two recent debris flows in Brazil. J S Am Earth Sci 113:103675
Dolliver HA (2012) Using Google Earth to teach geomorphology. Geol Soc Am Special Paper 492:419–429
Duszyński F, Ford DC, Goudie AS, Migoń P (2021) Rock properties and rock-controlled landforms. In: Burt TP, Goudie AS, Viles H (eds) The history of the study of landforms or the development of geomorphology. Volume 5: Geomorphology in the second half of the twentieth century. Geol Soc Lond Mem 58:151–171
Ezcurra E et al (2019) Natural experiment reveals the impact of hydroelectric dams on the estuaries of tropical rivers. Sci Adv 5(3):eaau9875
Fisher GB, Amos CB, Bookhagen B, Burbank DW, Godard V, Whitmeyer SJ (2012) Channel widths, landslides, faults, and beyond: The new world order of high-spatial resolution Google Earth imagery in the study of earth surface processes. Geol Soc Am Special Paper 492:1–22
Gabrielli S, Spagnolo M, De Siena L (2020) Geomorphology and surface geology of Mount St. Helens Volcano. J Maps 16(2):585–594
Gerrard AJ (1988) Rocks and landforms. Unwin Hyman, London, p 320
Gerrard AJ (2008) Geology and landforms. In: Burt TP, Chorley RJ, Brunsden D, Cox NJ, Goudie AS (eds) The history of the study of landforms, vol 4. Geological Society, London, pp 13–54
Goudie AM (2013) Characterising the distribution and morphology of creeks and pans on salt marshes in England and Wales using Google Earth. Estuarine Coastal Shelf Sci 129:112–123
Goudie AS (2016) Quantification of rock control in geomorphology. Earth Sci Rev 159:374–387
Goudie AS (2020) Global barchans: a distributional analysis. Aeol Res 44:100591
Goudie AS (2022a) Desert landscapes of the world with Google Earth. Springer, Cham, p 271
Goudie AS (2022b) Nebkhas: an essay in aeolian biogeomorphology. Aeol Res 54:100772
Goudie AS (2023a) Linear dunes: morphology from Google Earth. Desert 28(2):163–180
Goudie AS (2023b) Landscapes of the anthropocene with Google Earth. Springer, Cham, p 249
Goudie AS (2023c) Zibars: distribution, morphology and environmental controls. Aeol Res 60:100854
Goudie AS, Goudie AM, Viles HA (2021a) The distribution and nature of star dunes: a global analysis. Aeol Res 50:100685
Goudie AS, Goudie AM, Viles HA (2021b) Dome dunes: distribution and morphology. Aeol Res 51. https://doi.org/10.1016/j.aeolia.2021.100713
Harrison S et al (2024) Rock glacier distribution across the Himalaya. Glob Planet Change 239:104481
Hartmann J, Moosdorf N (2012) The new global lithological map database GLiM: a representation of rock properties at the Earth surface. Geochem Geophys Geosyst 13(12). https://doi.org/10.1029/2012GC004370
Henshaw AJ, Sekarsari PW, Zolezzi G, Gurnell AM (2020) Google Earth as a data source for investigating river forms and processes: Discriminating river types using form-based process indicators. Earth Surf Proc Landf 45(2):331–344
Jennings JN (1985) Karst geomorphology. Basil Blackwell, Oxford, p 304
Jonathan M, Mjemah IC, Kimambo ON, Hamad AA (2024) A geospatial approach to delineate lineaments on the granite gneissic terrain of Mpwapwa District, Dodoma. Heliyon (in press), Central Tanzania
Karydas C, Panagos P (2020) Towards an assessment of the ephemeral gully erosion potential in Greece using Google Earth. Water 12(2):603
Keller EA, Pinter N (2002) Active tectonics, uplift, and landscape, 2nd ed. Prentice Hall, Upper Saddle River, New Jersey, p p362
Komatsu G, Feyzullayev AA (2024) Geomorphology of subaerial mud volcanoes in Azerbaijan: issues about edifice construction and degradation. Geomorphology 463:109352
Kotluri SK, Pandey P, Pandey AK (2024) A Google Earth and ArcGIS-based protocol for channel width extraction. J Earth Syst Sci 133:article 9
Lageson DR, Larsen MC, Lynn HB, Treadway WA (2012) Applications of Google Earth Pro to fracture and fault studies of Laramide anticlines in the Rocky Mountain foreland. Geol Soc Am Spec Paper 492:209–220
Lee J (2015) Reconstruction of ancestral drainage patterns in an internally draining region, Fars Province. Iran. Geol Mag 152(5):830–843
Liang J, Gong J, Li W (2018) Applications and impacts of Google Earth: a decadal review (2006–2016). J Photogram Remote Sens 146:91–107
Lisle RJ (2006) Google Earth: a new geological resource. Geol Today 22(1):29–32
Liu X, Guo M, Chen Z, Zhang X, Yang F, Zhang S (2024) Quantifying the contributions of precipitation, topography and human activity and their coupling to the development of permanent gully. Geoderma 449:117015
Lobeck AK (1939) Geomorphology. McGraw-Hill, New York, p 731
Luo J, Niu F, Lin Z, Liu M, Yin G (2019) Recent acceleration of thaw slumping in permafrost terrain of Qinghai-Tibet Plateau: an example from the Beiluhe Region. Geomorphology 341:79–85
Ma L et al (2024) Factors influencing ephemeral gullies at a regional scale: formation and density. Land 13(4):553
Migoń P (2006) Granite landscapes of the world. Oxford University Press, Oxford, p 412
Migoń P (2020) Geomorphology of conglomerate terrains—global overview. Earth Sci Rev 208:103302
Migoń P (2025, in press) Rock controlled landscapes. Springer, Cham
Miller TG (1953) Geology and scenery in Britain. Batsford, London, p 224
Nainwal HC, Banerjee A, Shankar R, Semwal P, Sharma T (2016) Shrinkage of Satopanth and Bhagirath Kharak glaciers, India, from 1936 to 2013. Ann Glaciol 57:131–139
Ortiz-Guerrero C, Loizzo J (2024) Rocks Really Rock: electronic field trips via Web Google Earth can generate positive impacts in attitudes toward Earth sciences in middle-and high-school students. Geosci Commun 7(2):101–119

Pandey P (2019) Inventory of rock glaciers in Himachal Himalaya, India using high-resolution Google Earth imagery. Geomorphology 340:103–115

Paron P, Goudie AS (2007) Preliminary results about mapping and geomorphological correlation of tiger bush (Brousse tigrée) in Somalia, from a remote sensing and GIS analysis perspective. In: Goudie AS, Kalvoda J (eds) Geomorphological variations. Nakladatelství P3K, Prague, pp 37–86

Peng D, Yueren X, Qinjian T, Wenqiao L (2021) Using Google Earth images to extract dense landslides induced by historical earthquakes at the southwest of Ordos. China. Front Earth Sci 8:633342

Pereira MA, de Souza Alves A, Kobiyama M (2024) Determination and characterization of the junction angles in river channels in Brazil with Google Earth Pro. Rev Brasil Geomorfol 25(2). https://doi.org/10.20502/rbg.v24i2.2437

Rabby YW, Li Y (2019) An integrated approach to map landslides in Chittagong Hilly areas, Bangladesh, using Google Earth and field mapping. Landslides 16(3):633–645

Reece DA, Lory JA, Haithcoat TL, Gelder BK, Cruse R (2023) Using Google Earth imagery to target sssessments of ephemeral gully erosion. J Am Soc Agric Biol Eng 66(1):155–166

Samal AK, Srivastava RK, Ernst RE, Söderlund U (2019) Neoarchean-Mesoproterozoic mafic dyke swarms of the Indian Shield mapped using Google Earth™ images and ArcGIS™, and links with large igneous provinces. In: Srivastava K, Ernst RE, Peng P (eds) Dyke swarms of the world: a modern perspective. Springer, Singapore, pp 335–390

Sarmiento GN, Sarmiento NV, Delrieux CA, Perillo GM (2020) Morphological characterization of ponds and tidal courses in coastal wetlands using Google Earth imagery. Estuarine Coastal Shelf Sci 246:107041

Sati SP et al (2022) Mountain highway stability threading on the fragile terrain of upper Ganga catchment (Uttarakhand Himalaya), India. J Mountain Sci 19(12):3407–3425

Sawant S et al (2024) Integration of machine learning and remote sensing for assessing the change detection of mangrove forests along the Mumbai coast. J Earth Syst Sci 133(4). https://doi.org/10.1007/s12040-024-02378-0

Scheffers AM, Kelletat DH (2016) Lakes of the world with Google Earth: understanding our environment. Springer, Dordrecht, p 306

Scheffers AM, Scheffers SR, Kelletat D (2012) The coastlines of the world with Google Earth: understanding our environment. Springer, Dordrecht, p 305

Scheffers AM, May SM, Kelletat DH (2015) Landforms of the world with Google Earth: understanding our environment. Springer, Dordrecht, p 401

Schmid MO et al (2015) Assessment of permafrost distribution maps in the Hindu Kush Himalayan region using rock glaciers mapped in Google Earth. Cryosphere 9(6):2089–2099

Schumm SA (1956) The role of creep and rainwash on the retreat of badland slopes (South Dakota) Am J Sci 254:693–706

Scoon RN (2021) The geotraveller. Springer, Cham, p 398

Selby MJ (1993) Hillslope materials and processes, 2nd ed. Oxford Univ Press, Oxford, p 466

Shirani K, Peyrowan, Shadfar S (2023) Gully erosion mapping based on hydro-geomorphometric factors and geographic information system. Environ Monit Assess 195. https://doi.org/10.1007/s10661-023-11197-7

Sparks BW (1971) Rocks and relief. Longman, London p, p 422

Stamp LD (1946) Britain's structure and scenery. Collins, London, p 255

Sun H, Li W, Scaioni M, Fu J, Guo X, Gao J (2023) Influence of spatial heterogeneity on landslide susceptibility in the transboundary area of the Himalayas. Geomorphology 433:108723

Sunamura T (1992) Geomorphology of rocky coasts. Wiley, Chichester, p 302

Swanger W, Admassu Y (2018) Using Google Earth and Google street view to rate rock slope hazards. Environ Eng Geosci 24(2):237–250

Sweeting MM (1972) Karst landforms. Macmillan, London, p 378

Tewksbury BJ, Dokmak AA, Tarabees EA, Mansour AS (2012) Google Earth and geologic research in remote regions of the developing world: an example from the Western Desert of Egypt. Geol Soc Am Special Paper 492:23–36

Tooth S (2015) Google Earth as a resource. Geography 100:51–56

Traganos D, Aggarwal B, Poursanidis D, Topouzelis K, Chrysoulakis N, Reinartz P (2018) Towards global-scale seagrass mapping and monitoring using Sentinel-2 on Google Earth engine: the case study of the Aegean and Ionian seas. Remote Sens 10(8):1227

Trenhaile AS (1987) The geomorphology of rocky coasts. Clarendon Press, Oxford, p 384

Twidale CR, Romaní JRV (2005) Landforms and geology of granite terrains. Balkema, Leiden, p 351

Warnasuriya TW, Kumara MP, Gunasekara SS, Gunaalan K, Jayathilaka RM (2020) An improved method to detect shoreline changes in small-scale beaches using Google Earth Pro. Marine Geol 43(6):541–572

Whalley WB (1976) Properties of materials in geomorphological explanation. Oxford University Press, Oxford, p 60

Whalley WB (2021) Geomorphological information mapping of debris-covered ice landforms using Google Earth: an example from the Pico de Posets, Spanish pyrenees. Geomorphology 393:107948

Whittow JB (1992) Geology and scenery in Britain. Chapman and Hall, London, p 478

Xu LR, Su ZM, Ugai K, Cai F (2011) Yang QQ (2011) Google Earth as a tool in terrain survey of debris flow watershed. Adv Materials Res 261:1563–1566

Yang HA, Weitao YU, Zhongping LA, Shixiu CH, Wenke LI (2023) Extraction of yardang landforms boundary based on multi-spatial resolution Google Earth image and Canny edge algorithm. Bull Surv Mapping 25(5):175–179

Yatsu E (1966) Rock control in geomorphology. Sozosha, Tokyo, p 135

Young RW Wray RLA Young ARM (2009) Sandstone landforms. Cambridge University Press, Cambridge, p 31

Structures

2

Abstract

Geological structures, such as joints, folds (including anticlines, monoclines and synclines), faults, and domes, are highly important in determining the role that rocks play in landform development. Joints are a type of rock fracture, which unlike a fault, show no evidence of movement on either side. Joints can be formed by a number of factors, including tectonic stresses, changes in the temperature of a rock mass, and release of pressure by erosion of overlying rocks (unloading). Folds are structures that originally were planar, like a sedimentary bed, but which have been bent by horizontal or vertical forces in Earth's crust. Folds can occur at scales that range from mountain ranges to small crumples only a few centimetres long. Faults are breaks in the surface rocks of the earth's crust as a result of sudden yielding under unequal stresses. Faulting is accompanied by displacement along a fault plane. There are also many different types of igneous masses. Some are very large and are known to have sizes up to 1600 × 160 km. The scale can decrease down to that of a volcanic plug eroded out from its surrounding ash beds. Some are intrusive and some are extrusive. The former are rocks which cool and solidify within the crust, whereas the latter are those which are poured out on to Earth's surface. Among the consequences of intrusion are domes, dykes and ring structures. Some other distinctive structures are associated with rocks composed of salt (halite).

Keywords

Fault · Fold · Thrust · Joint · Intrusions · Doming · Anticline · Syncline · Salt

2.1 Introduction

Geological structures are a very fundamental control of Earth's geomorphology, inclduding the forms of escarpments and the nature of mass movements (Duszyński et al. 2019). Landforms and their disposition are associated with such factors as jointing, folding, faulting, igneous intrusion and extrusion, and salt doming. Google Earth can be very useful in identifying and mapping geological structures (see, for example, Lageson et al. 2012; Lawal et al. 2022). Anhaeusser (2024) identified previously unknown structures in greenstone terrains of the Kruger National Park in South Africa, while Carr (2017) used Google Earth to produce an inventory of volcanoes in central America. Google Earth has also been used to make inventories of dyke swarms (Samal et al. 2019).

2.2 Joints

A joint is a type of rock fracture, which unlike a fault, shows no evidence of movement on either side. Joints can be formed by a number of factors, including tectonic stresses, changes in the temperature of a rock mass, and release of pressure by erosion of overlying rocks (unloading). They occur in a wide range of rock types including sandstones, as in Wadi Rum, Jordan (see Sect. 5.2.2) and Arches, USA (see Sect. 5.2.7), in granites, as at Joshua Tree, USA (see Sect. 3.4.4), and in basalts, as at Victoria Falls (see Sect. 4.7). In northern Australia (Fig. 2.1) there are large tracts of ancient terrain in rocks that may be c 2 billion years old, that display a hierarchy of joints. The same is true of a limestone area further south in the Devonian Oscar Range in the Kimberley District, where the terrain shows a series of platforms and pediments with upstanding hillocks that are strongly joint-controlled. Similarly, in the so called giant grikeland of the 'lost city' of the Barkly Tableland there is remarkable control of karst features by a network of joints (best seen at c 18 41° S, 138 20° E) (Gale et al. 1997).

A. Goudie, *Rocky Landscapes of the World with Google Earth*, https://doi.org/10.1007/978-3-031-77418-8_2

(a)

(b)

Fig. 2.1 **a** Joint systems in ancient rocks in the north of the Northern Territory, Australia. 13° 8′ 0.98″ S, 133° 13′ 44.83″ E. ©Google Earth. Scale bar = 4 km. **b** Joint-controlled limestone terrain in the Oscar Range, Kimberley, Western Australia. 17° 58′ 44.78″ S, 125° 23′ 8.28″ E. ©Google Earth. Scale bar = 50 m

Fig. 2.2 The great fjords of southern Chile, showing their alignments and angularities associated with various structural controls. 51° 34′ 34.20″ S, 73° 34′ 16.12″ W. ©Google Earth. Scale bar = 80 km

Joints can act as avenues for weathering and erosion, and can also control the density and orientations of phenomena such as stream segments. They can determine rates of glacial quarrying (Hooyer et al. 2012), the nature of landsliding (Cody et al. 2020), and the form of crenulations in escarpments (Nicholas and Dixon 1986). Measurements of joint spacing, width, continuity and orientation are an important component of rock mass strength classifications, and play a major role in determining escarpment slope forms (Moon and Selby 1983) and the form of inselbergs (Selby 1982).

Glacial fjords (fiords) often show a series of clear alignments and angular courses associated with joints, fractures and other discontinuities. This is evident both in Norway (Nilsen 1973) and in southern Chile (Fig. 2.2). Expanded Pleistocene glaciers excavated along these lines of weakness. Glasser and Ghiglione (2009, p. 300) made the following points with regard to southern Chile.

> 1. The presence of a pervasive structural network of lineaments with strong preferred orientations was conditioned by a long-lived history of wrench kinematics that affected the region. 2. There is a strong correspondence between mapped structural geological lineaments and fjord orientation throughout the southern Andes. 3. The underlying geological structure exerted a very strong control on the pattern of ice discharge, and therefore constraining the pattern of fjord development beneath the former Patagonian Ice Sheets. 4. The location of fast-flowing outlet glaciers from the former Patagonian Ice Sheets was determined more by the location of these pre-existing structural geological weaknesses than by glaciological controls on patterns of ice flow.

One earlier researcher on fjords, Gregory (1913), went so far as to suggest that fjords largely had a preglacial origin, were substantially moulded by fluvial processes, that glaciers had only played a limited role in their formation, and that they were largely of tectonic origin (i.e. that they were structural features). There are grains of truth amid some of his bolder assertions, but many of his ideas are now discounted.

Other glacial features, including cirques and lake basins, may show the effects of structures and lithology and a penetrating analysis of this complex issue is provided by Evans (1994). In NW Scotland, rock jointing and hardness seem to have controlled the form of glacial valleys and Krabbendam and Glasser (2011, p. 374) compared two rock types and found that:

> Torridon sandstone is soft but thick-bedded and with a wide joint spacing. Erosional bedforms include roche moutonnées with smoothed tops and concave stoss sides, whalebacks, and elongate p-forms, indicating a high proportion of abrasion over plucking. Cambrian quartzite is hard but thin-bedded with narrow joint spacing. Erosional landforms are angular to sub-angular with abundant plucked lee faces, suggesting a high proportion of plucking over abrasion. Hardness and joint spacing thus exert a strong control on subglacial erosional landforms and the mechanisms that formed them

Absence of joints is also a very significant control of some rocky landscapes, not least those in granite regions, where the size and shape of inselbergs is largely controlled by the presence or absence of joints (see Sect. 3.4). Figure 2.3a, b

(a)

(b)

Fig. 2.3 **a** Mount Amy, Kimberley, Western Australia. 17° 9′ 20.27″ S, 124° 53′ 11.41″ E. ©Google Earth. Scale bar = 200 m. **b** Mount Amy, Kimberley, Australia (*Photo* A. S. Goudie)

shows an inselberg, Mount Amy, from the West Kimberley District of Western Australia. Large compartments of granite from the palaeoproterozoic (c 1800 mya) are without joints and this gives rise to the impressive domes rather than to smaller castle koppies, tors, and boulders. By contrast, some granite landforms are controlled by closely spaced joints, as with some outcrops in NE Sardinia (Fig. 2.4a, b).

(a)

(b)

Fig. 2.4 **a** Densely jointed Permo-Carboniferous granites on Caprera Island, NE Sardinia. 41° 13′ 45.23″ N, 9° 27′ 35.14″ E. ©Google Earth. Scale bar = 80 m. **b** Permo-Carboniferous granites with a dense network of vertical and hozizontal joints at Porto Cervo, NE Sardinia (*Photo* A. S. Goudie)

2.3 Folds and Thrusts

Folds are structures that originally were planar, like a sedimentary bed, but which have been bent by horizontal or vertical forces in the Earth's crust. Fold mountains are the most common type of mountain. They result from compressional tectonics at convergent plate boundaries. When tectonic plates collide, the leading edges of continental crust crumple and fold. This compression and uplift process leads to the formation of massive, mountain ranges, typically long, linear ranges that stretch for hundreds or even thousands of kilometers. This linear shape reflects the direction of the compressive forces acting along the plate boundary. The Himalayas, Andes (Fig. 2.5), and Alps are all active fold mountains, though in most cases it is both folding and thrusting that formed them.

This is certainly the case for the great Karakorams of Pakistan and its neighbours (Goudie et al. 1984; Searle 1991, 2013) (Fig. 2.6). This is Earth's second highest mountain range, and is part of a complex of ranges that includes the Pamirs, Hindu Kush and Himalayas. It contains 18 summits higher than 7500 m (with four above 8000 m) and is the result of Indian–Asian continental–continental collision. It is currently uplifting and is being intensely denuded, not least by large glaciers, producing some of the greatest relative relief on Earth. The region contains some of the largest rivers with the highest sediment loads in the world, including the Indus, Gilgit and Hunza. It has a very complex history with major thrusts and faults, batholith intrusions, island arc sedimentation and metamorphic zones. The Karakoram Metamorphic Complex (KMC) in the southern Karakoram block is an exceptionally fine example of Barrovian type metamorphism that comprises numerous exhumed metapelite units. This complex is bounded to the north by the granitoids of the massive east–west trending Karakoram axial batholith and to the southwest by the Main Karakoram Thrust (MKT) (Baig et al. 2020).

The Southern Alps in New Zealand (Fig. 2.7) are another example of the importance of thrusting. They run approximately 500 km northeast to southwest. Their tallest peak is Mount Cook (Aoraki) at 3724 m. The Southern Alps lie along a plate boundary, with the Pacific Plate to the southeast pushing westward and colliding with the northward-moving Indo/Australian Plate to the northwest. Over the last 45 million years, the collision has pushed up a 20 km thickness of rocks on the Pacific Plate to form the Alps, although much of this has been eroded away because of

Fig. 2.5 Peaks in the Andes of northern Chile, inland from Arica. The Andes are the result of the subduction of oceanic crust beneath the South American Plate as the Nazca Plate and the South American Plate converged (*Photo* A. S. Goudie). Many high volcanoes occur

(a)

(b)

Fig. 2.6 **a** The Karakoram Mountains and Hunza River, Pakistan. 36° 31′ 7.91″ N, 74° 53′ 52.04″ E. ©Google Earth. Scale bar = 5 km. These are among the highest mountains in the world and were formed by both thrusting and folding associated with the northward movement of the Indian continental plate into the southern margin of Asia. **b** The Karakoram Mountains near Passu in the Hunza Valley in Pakistan (*Photo* A. S. Goudie). Note the screes and debris fans on the far side of the river. These are the same ones shown in the centre of the above satellite image

(a)

(b)

Fig. 2.7 **a** The Southern Alps of New Zealand, showing the Alpine Fault. 43° 0′ 37.41″ S, 171° 20′ 18.40″ E. ©Google Earth. Scale bar = 50 km. **b** The Southern Alps of New Zealand at Arthur's Pass (*Photo* A. S. Goudie)

high rainfall, glaciation and high relief. The plate boundary is marked by the Alpine Fault, a remarkably straight feature clearly visible on Google Earth. Significant crustal compression has occurred and as Summerfield (1991, p. 76) remarked, 'The western slopes of the Southern Alps are, in effect, an enormous self-pepetuating fault scar; as rapidly as material is eroded away it is replaced by further uplift and crustal shortening'. In some parts uplift exceeds 7 mm per annum, but rates of erosion are also very high and provide the bedload of the multi-thread rivers flowing eastwards to the Canterbury Plains.

In South Africa there are the Cape Fold Mountains (Fig. 2.8a). These consist of a series of parallel ranges that run along the south-western and southern coastlines of South Africa for 850 km from the Cederbergs (Fig. 2.8b) (200 km to the north of the Cape Peninsula), and then along the south coast as far as Gqeberha (Port Elizabeth), 650 km to the east. The Cape Fold Belt (CFB) is believed to have formed in response to subduction-related compression during Paleozoic–Mesozoic convergence along the south-western margin of Gondwanaland (Manjoro 2015). It could equally well be called the Cape Thrust Belt, for both thrusting and folding contributed to the formation of this region.

The CFB is part of a large orogenic belt, which prior to the breakup of Gondwanaland, extended west into South America, and east through the Falkland Islands and into Antarctica. The CFB is mostly made up of Ordovician to Carboniferous continental margin clastic sedimentary rocks of the Cape Supergroup, which comprises three lithostratigraphic groups: the Table Mountain, Bokkeveld and Witteberg Groups. The first of these consists mostly of medium-grained quartzites, shale and sandstone, while the rocks of the Bokkeveld Group are predominantly shales with a small proportion of sandstones and quartzites. On the other hand the Witteberg Group is predominantly composed of quartzitic sandstone and shale. The mountains have synclinal valleys, often developed in the shales, and ridges built of the sandstones and quartzites. In spite of the high relief and steep slopes, cosmogenic dating has suggested that rates of denudation are low and Scharf et al. (2013) attributed the maintenance of rugged topography and the suppression of denudation rates primarily to the presence of the physically robust and chemically inert quartzites that constitute the backbone of the mountains.

The Ural Mountains form a natural border between Europe and Asia, stretching c 2500 km southwards from the Arctic Ocean to the Ural River. They are generally classified as a fold mountain range, formed primarily by compressional forces associated with a convergent boundary between the Siberian and Kazakh (Eurasian) continental plates (https://geoscience.blog/the-formation-of-the-ural-mountains-unveiling-earths-geomorphological-marvels/) (Accessed 14th August 2024). The mountains owe their existence to a major tectonic event known as the Ural Orogeny, which began about 300 million years ago during the Late Paleozoic. This involved the subduction of the Siberian plate beneath the Eurasian plate, the compression and folding of rocks, and the uplift caused by the collision between the two plates. Additionally, volcanic activity and the intrusion of igneous rocks played a role in shaping the mountain range. In spite of their age, which is much greater than that of the Karakorams, Andes or the New Zealand Alps, the Urals still have impressive relief (Fig. 2.9), with the highest point being Mount Narodnaya, which reaches a height of 1,894 m.

Other ancient fold landscapes occur in northeastern Canada in an area generally referred to as the New Quebec Orogen or the Labrador Trough (Fig. 2.10). In this case, during the Pleistocene, advancing and retreating glaciers scraped the surface clear of debris to help make visible some beautiful fold patterns developed in flood basalts (see Sect. 4.7 for further examples of these). These were laid down about 2.17 billion years ago. The gentle folding of the basalts, to give the Corrugated Hills, came later with the collisions of cratons about 1.87 to 1.80 billion years ago (van Rooyen et al. 2024).

The great inlets of the sea—rias—adjacent to Talbot Bay in the Kimberley district of northwestern Australia, are another important instance of ancient folding. In this case the area's sandstone beds, which formed in the c 1840 million year old Warton and Pentecost Sandstones, are almost vertical (upto 85°), and form spines that arise from the sea. Some of these spines have been breached, providing routes for tidal flows to produce what are called 'horizontal waterfalls' (Fig. 2.11)—*Garaanngaddim* by the local aboriginal people. The twin breaches are part of the McLarty Ranges, which have two ridges running parallel approximately 300 m apart. The first and most seaward gap, through the Pentecost Sandstone, is c 20 m wide and the second, through the Warton Sandstone, is a mere 10 m wide. The phenomenon is caused by the changing local sea level due to tides of up to 10 m in height. The water builds up on one side or the other of the narrow gaps faster than it can flow through them, creating a waterfall up to 4 m high. With each change of the tide, the direction of the falls reverses from inflow to outflow and back again (Scott 2015).

Folds can occur at scales that range from great mountain ranges, such as those just mentioned, to small crumples only a few centimetres long. They can be gentle or severe (Fig. 2.11a), depending on the nature and magnitude of the applied forces and the ability of the beds to be deformed. When beds are up-folded or arched they are called *anticlines*, whereas down-folds or troughs are termed *synclines*. A *monocline* is a step-like bend in otherwise horizontal or gently dipping beds. A *pericline* is a fold shown on a geological map as concentrically arranged contour patterns

(a)

(b)

Fig. 2.8 **a** Part of the Cape Fold (Thrust) Mountains, South Africa. 33° 36′ 2.20″ S, 22° 10′ 3.40″ E. ©Google Earth. Scale bar = 100 km. **b** Dipping strata exposed in the Cederberg Mountains, Western Cape, South Africa. The Cederberg Mountains are well described by Quick and Eckardt (2015). They are part of the Cape Fold Mountains (*Photo* Alice Goudie)

(a)

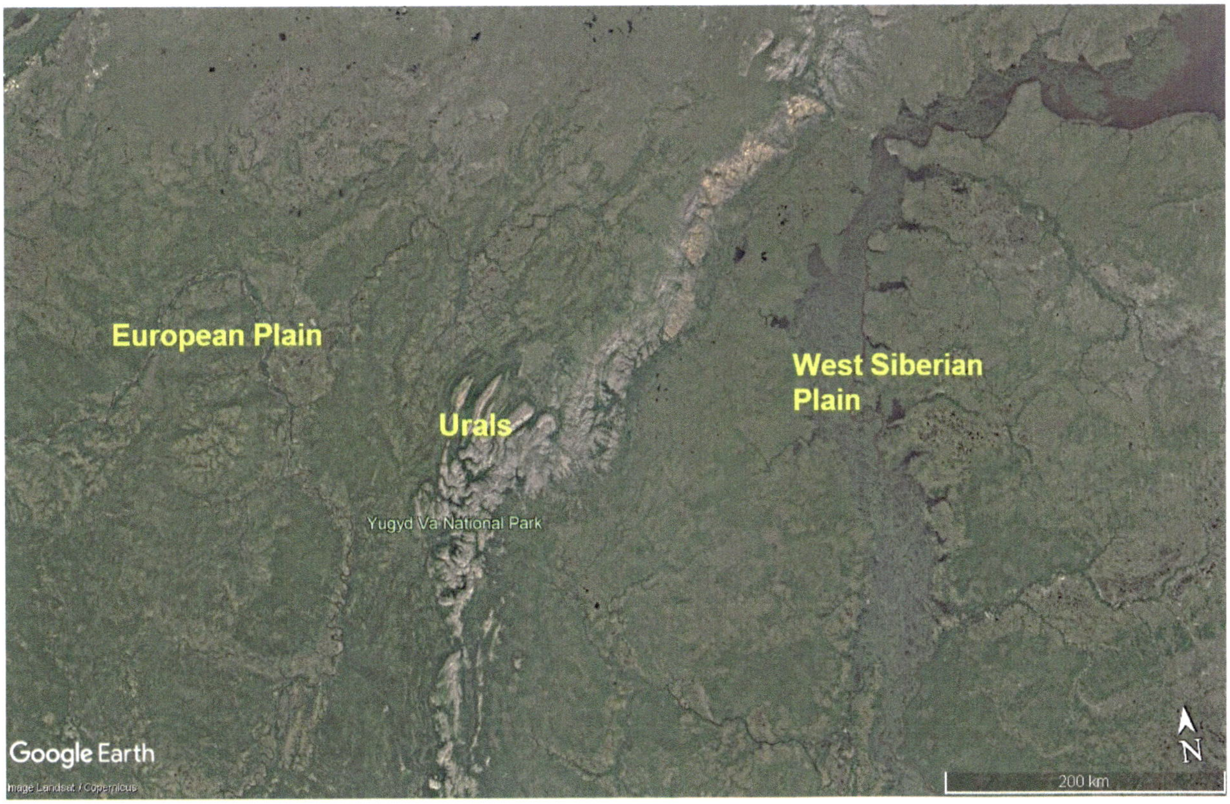

(b)

Fig. 2.9 **a** The Ural Mountains, Russia. 65° 21′ 13.81″ N, 61° 54′ 43.74″ E. ©Google Earth. Scale bar = 200 km. **b** The Ural Mountains in the Yugdv Va National Park, Russia. *Source* https://en.wikipedia.org/wiki/Yugyd_Va_National_Park. Public Domain

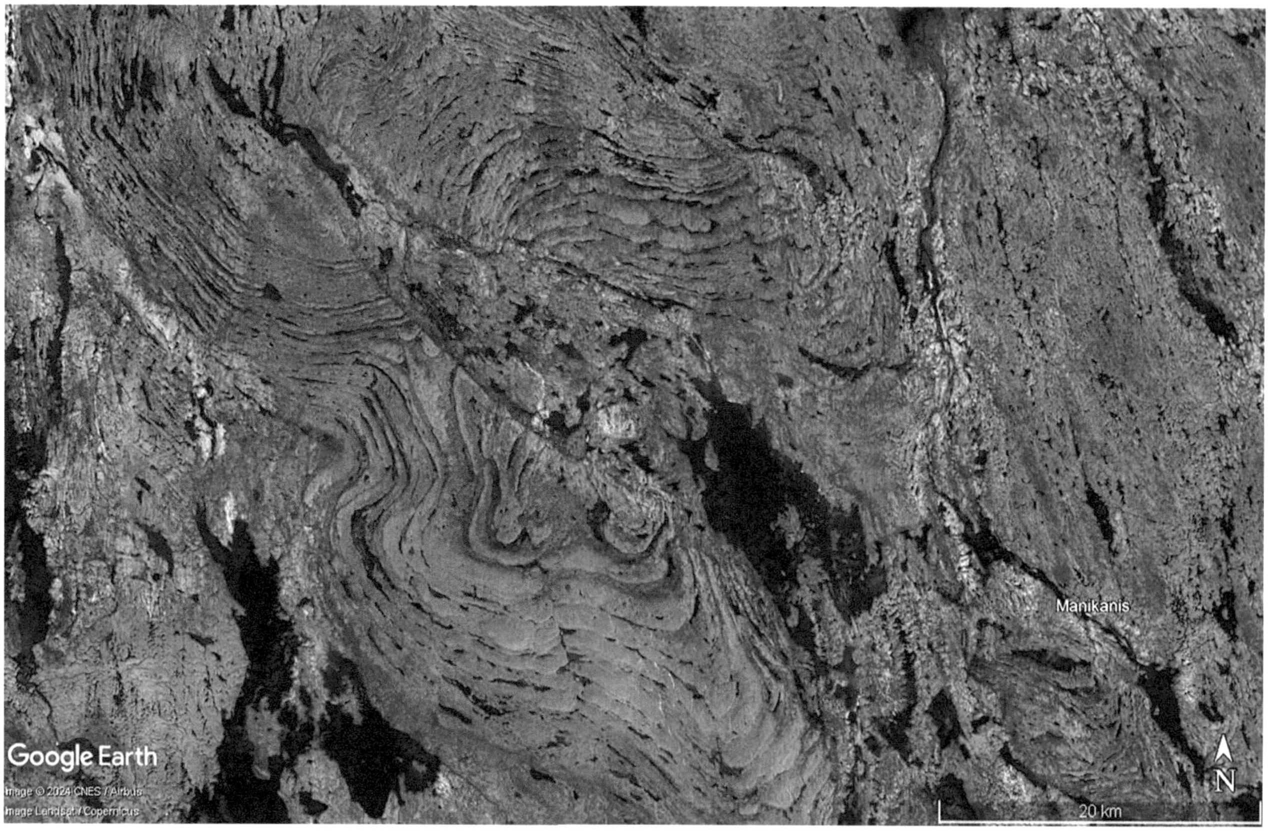

Fig. 2.10 Folded flood basalts in the Corrugated Hills of the New Quebec Orogen, Canada. 56° 19′ 19.14″ N, 67° 51′ 23.16″ W. ©Google Earth. Scale bar = 20 km

whose amplitude decreases to zero in a regular fashion in two directions. An anticlinal pericline is called a 'dome', and a synclinal pericline is called a 'basin'. In some cases, including the Alps, folding and thrusting cause strata to be formed into *nappes,* a *nappe* being a package of rock that has been folded and deformed to the point that it is completely allocthonous, that is disconnected from its origin. Compressive stresses associated with convergent plate tectonic settings sometimes manage to fold beds into up-side-down positions. Strata may then be doubled up over each other (Fig. 2.12b). Sometimes folds may be rather angular in form, and these are called *chevron* folds.

Along the northern coast of the Sultanate of Oman and the eastern portion of the United Arab Emirates, in the Hajar Mountains, are a series of nappes associated with what is known as an ophiolite belt. This refers to an intact section of oceanic lithosphere that has been thrust onto a continental margin. The Samail (Semail) Ophiolite, is one of the world's largest and best-exposed examples. Rocks include ultramafic peridotite (harzburgite and dunite) and gabbro. Ophiolite originated at c 96–94 mya at a spreading centre above a northeast-dipping subduction zone c 500 km long and 50–100 km wide (Searle and Cox 1999; Rollinson 2014). The ophiolite forms much of the mountainous terrain of the area (Fig. 2.13). Within these mountains are sections of the deep Earth, rarely exposed anywhere else – pieces of the Earth's mantle and the lower parts of its oceanic crust. The ophiolite rocks are subject to weathering and liberate carbonates that form tufas. Many alluvial fans have been deposited as a result of erosion (Woor et al. 2023).

The Troodos Mountains of Cyprus are also noted for their ophiolites created by the collision of the African and Eurasian plates, while in New Caledonia ophiolite is on the main island and was formed by obduction of the Loyalty oceanic basin onto the continental crust of the Norfolk Ridge during the late Eocene (c 34 mya). Karstic weathering phenomena associated with its peridotites are discussed in Sect. 6.6.

As regards folds, initially, an anticline will form an area of upstanding, often domal relief, as with the folds in the Zagros Mountains of Iran (Fig. 2.14). However, as it is eroded, the uppermost strata may be cut through. If the older rock that forms the core is of low resistance the result is a *breached anticline* in which a series of inward-facing escarpments rise above a central lowland, as with the Mahktesh landforms of Israel. Conversely, the rim of a syncline will possess outward facing scarps. The angle of dip also influences topographic form. Resistant beds in

(a)

(b)

Fig. 2.11 **a** The Horizontal Falls of the Kimberley coast, Australia. 16° 22′ 51.66″ S, 123° 57′ 35.42″ E. ©Google Earth. Scale bar = 2 km. **b** The Horizontal Falls, Talbot Bay, Kimberley, Australia, showing the gaps cut through steeply dipping sandstones. The swirling water of tidal flow is shown in the foreground (*Photo* A. S. Goudie)

Fig. 2.12 **a** Steeply-dipping folded strata in Ras Al Khaimah, UAE (*Photo* A. S. Goudie). **b** Some major types of fold (modified from https://www.pmfias.com/fold-fault-block-residual-mountains/) (Accessed 28th June 2024)

very gently dipping or horizontal beds, form flat-topped plateaux (*mesas* and *buttes*). Modestly dipping beds create a *cuesta* or *escarpment*, whereas steeply dipping strata produce a ridge known as a *hogback*. Nearly vertical strata produce *razorbacks*. A good guide to using Google Earth to examine fold structures is https://serc.carleton.edu/NAGTWorkshops/structure/google_earth_mapping_locations.html (Accessed 3rd May 2024).

Folding influences mass movements in sedimentary rocks. If the bedding tilts away from the rock face, there may be a tendency for translational slides to develop, whereas if the bedding tilts into the rock face, there may be

(a)

(b)

Fig. 2.13 **a** The dark rocks of the mountainous Oman ophiolite belt. 23° 7′ 42.26″ N, 58° 12′ 40.36″ E. ©Google Earth. Scale bar = 80 km. The ophiolite rocks of Oman, with an area of carbonate deposition in the foreground (*Photo* A. S. Goudie)

Fig. 2.14 Iran salt domes and anticlines. 26° 52′ 51.95″ N, 54° 30′ 35.77″ E. ©Google Earth. Scale bar = 50 km. D = domes

a tendency for rotational slips to occur. This is evident on Portland on the south coast of England, where well-jointed but massive Jurassic Portland limestones overlie less competent Portland Sands, Portland Clays and Kimmeridge Clay. On the west side of the island, part of the Jurassic Coast, the bedding is inclined into the cliffs and a series of slips and topples have developed (Fig. 2.15a, b), forming arcs in the sea beneath. However, on the east wide of the island the beds dip out from the cliffs and here the style of landslipping is very different. Erosion at the base has caused great slices of limestone to topple and slide forward, with a 60–75 m vertical displacement. Upto six slices form an incredibly rough foot-slope around the cliff, and large blocks of strata display steeply dipping bedding (Fig. 2.15c, d) (Brunsden et al. 1996). The slide took place on February 3rd 1665 and involved the creation of an undercliff c 1.4 km long and up to 0.3 km wide. It was possibly the largest individual landslide event ever recorded in historical times in Britain. As Duszyński et al. (2019) have shown, where ductile rocks are overlain by a brittle and jointed unit, which is the case on Portland, extensive rock spreading may become a crucial geomorphic agent. Loading caused by the limestone caprock, together with the low shear strength and viscoplastic behaviour of the underlying clay layers, result in the deformation of the latter and the initiation of movement of caprock blocks on cliff or escarpment edges.

In the Negev Desert of Israel there are some structures called *Makhteshim*, which consist of anticlines that have been breached by erosion (Krasnov and Mazor 2001; Finzi and Harlev 2016). They are sometimes referred to as erosion cirques (Zilberman 2024), but this is not a useful term, for although erosional, they are not remotely like cirques. A rather better term, used by some Israeli scientists, is pseudo-crater.

Makhtesh Ramon is located c 85 km south of the city of Beersheba. The formation is 40 km long, 2–10 km wide and 500 m deep, and is shaped like an elongated heart. Rather smaller is Mahktesh Gadol (Fig. 2.16), which is shaped like an eye. Where a hard outer layer of rock covers softer rocks, erosion removes the softer minerals relatively quickly, and they are washed away from under the harder rock. The hard rocks are limestone and dolomites, while the inner softer rocks are chalk or sandstone. Over 200 mya (million years ago), the area was under the Tethys Sea. As this sea receded, it revealed a small anticlinal hill that had previously been submerged. This became exposed to the atmosphere, and rivers excavated the inside of the hill—which was comprised of soft limestone—to create

(a)

(b)

Fig. 2.15 **a** The cliffs on the west side of Portland, southern England, showing the cliff line and an arc-like mass of the toe of a large rotational slip. ©Google Earth. 50° 33′ 5.84″ N, 2° 26′ 59.84″ W. Scale bar = 100 m. **b** The cliffs on the west side of Portland, southern England, showing the cliff line and, in the background, an arc-like rotated mass of the toe of a large slip (*Photo* A. S. Goudie). **c** The great translational slide on the east side of Portland. 50° 32′ 32.59″ N, 2° 25′ 15.76″ W. ©Google Earth. Scale bar = 400 m. **d** View of the eastern cliffs of Portland from the sea. Note the cliff of Portland limestone in the background and the foundered and tilted blocks in the foreground. They came down from the cliff line in 1665 (*Photo* A. S. Goudie)

(c)

(d)

Fig. 2.15 (continued)

the crater-like form. Recently, the term Makhtesh has been applied to describe the structure of the Wealden Dome in south east England. Here the more resistant Chalk formation and underlying Greensands provide the rim of the North and South Downs, whereas the anticline's interior is lower-lying and exposes the older, underlying, clays and sandstones (Muir-Wood 2024, p. 249). The lowest parts of the dome are formed in clays—the Wealden Clays—and the highest parts at the heart of the dome are composed of sandstones of the Hastings Beds. Some of these give rise

Fig. 2.16 Makhtesh Gadol, Israel. 30° 55′ 54.53″ N, 34° 59′ 44.33″ E. ©Google Earth. Scale bar = 5 km

to valley-side cliffs and tors (Jones 2020; Duszyński and Migoń 2020).

In Utah, USA, is the San Rafael Swell (Hawley et al. 1968) (Fig. 2.17). This is a broad anticlinal flexure measuring approximately 121 by 64 km, and consists of a giant dome-shaped anticline of sandstone, shale, and limestone that was pushed up during the Paleocene Laramide Orogeny 60–40 mya. It was formed when deeply buried Precambrian basement rocks below the present-day Swell moved upwards relative to the surrounding areas and caused the overlying sedimentary rocks to form an anticline. The resulting structure has been likened to a series of blankets draped over a box. Resistant units that range in age from Permian to Early Tertiary age form a series of roughly elliptical inward-looking hogback ridges and strike valleys that define the fold. Navajo Sandstone forms the upper part of the bold cliff that defines the culmination of the structure. The Swell has a gently dipping western margin and a monoclinal eastern margin, where beds locally dip near 80° (https://storymaps.arcgis.com/stories/5a5705c963b54005820793a3a54bf01e) (Accessed 18th June, 2024).

A monocline is also well displayed in the eastern Dasht-e Lut desert in Iran. In a monocline there is a step-like fold in the strata consisting of a zone of steeper dip within an otherwise horizontal or gently dipping sequence (Fig. 2.18).

On the eastern side of the USA are the Appalachian Mountains (Fig. 2.19). These stretch northeastwards 3200 km from Alabama to Newfoundland. Here again the evidence of complex folding in the landscape is abundantly clear, dictating the ridge and valley network developed in Palaeozoic strata. The processes that led to the formation of these mountains started 1.1 bya, when the first mountain range in the region was created when the continents of Laurentia and Amazonia joined, creating a supercontinent called Rodinia. Many of the rocks and minerals that were formed during that event currently outcrop at the surface. Around 480 mya, geological events began that led to three distinct orogenic eras that created much of the surface structure seen in today's Appalachians. During this period, the Appalachians once reached elevations that were comparable to those of ranges like the Alps and the Rockies, before erosion occurred over the last 240 million years leading to the landscape we see now. The nature and timing of that erosion has been the subject of considerable debate (see Thornbury (1965, Chap. 5), and Morisawa (1989) for a discussion of the earlier theories).

The Finke River (*Larapinta* in the indigenous Arrernte language) in central Australia flows through an anticlinal structure in the Krichauff and James Ranges (Fig. 2.20) (Pickup et al. 1988). This river's course is over 300 my old

Fig. 2.17 San Rafael Swell, Utah, USA. 38° 54′ 5.64″ N, 110° 44′ 5.72″ W. ©Google Earth. Scale bar = 40 km

Fig. 2.18 Monocline in the Eastern Dasht-e Lut, Iran. 32° 24′ 40.36″ N, 56° 47′ 30.06″ E. ©Google Earth. Scale bar = 3 km

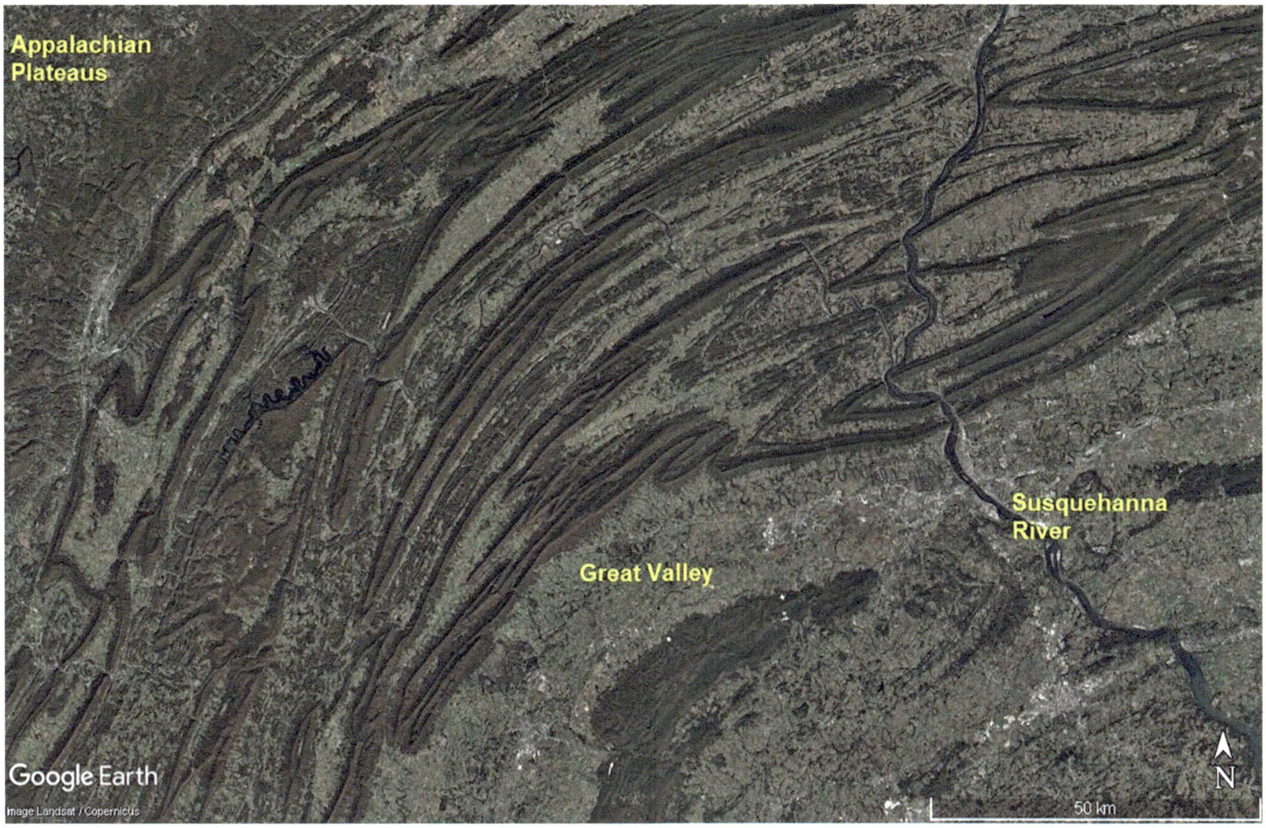

Fig. 2.19 Folds in the Appalachian Mountains, USA. 40° 20′ 40.49″ N, 77° 28′ 0.51″ W. ©Google Earth. Scale bar = 50 km. A structural and stratigraphic map of this area is given in Morisawa (1989, Fig. 12)

Fig. 2.20 Folds across the Finke River in central Australia. 24° 10′ 11.74″ S, 132° 52′ 48.98″ E. ©Google Earth. Scale bar = 8 km

Fig. 2.21 Structures in metamorphosed Neoproterozoic sedimentary rocks, Ugab Valley, Namibia. 20° 58′ N, 14° 02 E. ©Google Earth. Scale bar = 10 km

and it existed before the anticlinal structure was exhumed by uplift and denudation (Jansen et al. 2022). The river kept its meandering course and incised its valley in pace with the uplift. It is, therefore, an example of an antecedent river.

Metamorphic rocks, of which there are many different types, may also display structures. This is the case with the low-grade metamorphosed Neoproterozoic sedimentary rocks of the Ugab Valley in central Namibia (Fig. 2.21).

2.4 Faults

Faults are breaks in the surface rocks of the earth's crust as a result of sudden yielding under unequal stresses. Faulting is accompanied by displacement along a fault plane, as shown in the image of Piqiang in China (Fig. 2.22). This is a tear fault zone in the southern Tienshan Mountains (Li et al. 2020).

Faults take various forms that can be classified according to (i) the angle of inclination and (ii) the relative direction of the displacement. A *normal fault* has a steep or nearly vertical plane, and movement is predominantly in a vertical direction, with one side being raised (or upthrown) relative to the other (which is downthrown). In a *reverse fault* the beds on one side of a fault are thrust over those on the other side as a result of compressive forces. With *transcurrent* and *transform faults*, the displacement is horizontal along the line of the fault (Fig. 2.23).

In the case of *half-grabens,* they have subsided and rotated downward along a major *border fault*. These are single normal faults that bound just one side of a basin. The border faults are curved *(listric faults)*; this means they are steep at the surface but that they flatten at depth. This allows the basin to tilt downward toward the fault. Rift lakes lie in the depression near the border fault. African rift lakes include Malawi, Tanganyika, Turkana, and Kivu.

Faults and fractures influence relief in three main ways: by the displacements that they create, by the modification of the rocks in their vicinity (crush zones, breccias, etc.), and by the juxtaposition of rocks of differing weathering and erosion qualities. Among the landforms produced are fault scarps with distinctive facets, rivers with disrupted courses,

Fig. 2.22 The Piqiang Faults in China. ©Google Earth. 40° 17′ 13.04″ N, 77° 42′ 23.06″ E. Scale bar = 5 km. The upper reddish layers near the top of the sequence are Devonian sandstones formed by ancient rivers. The green layers in the middle are Silurian sandstones formed in a moderately-deep ocean. The cream-coloured layers at the base are Cambrian-Ordovician limestones formed in a shallow ocean

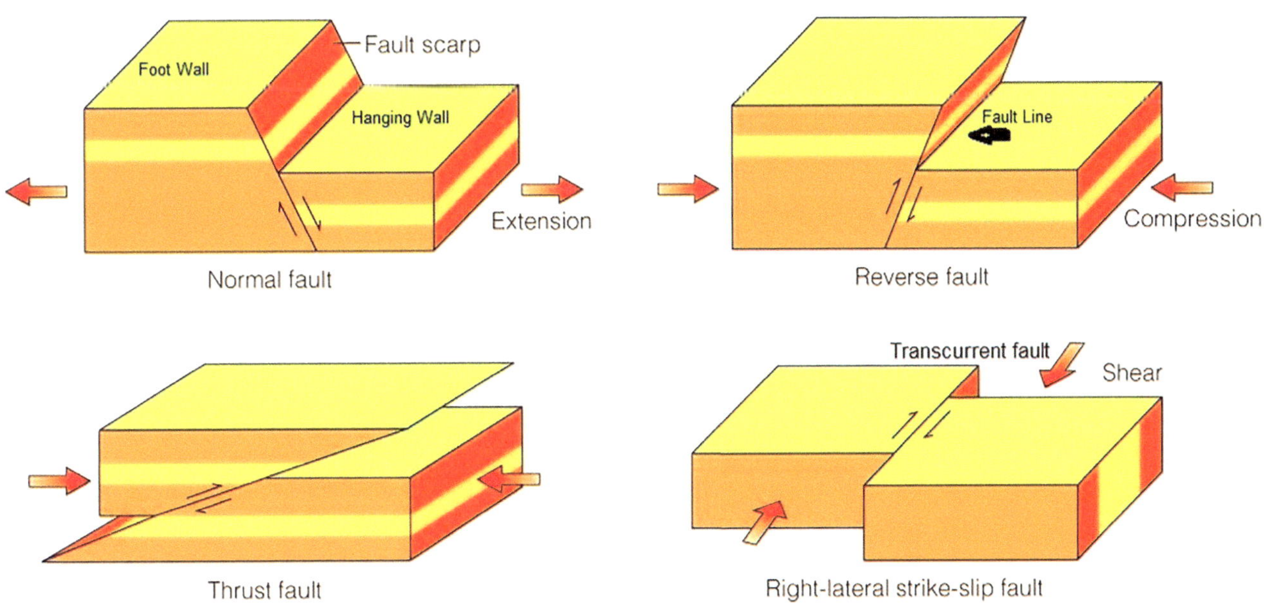

Fig. 2.23 Some major types of fault (modified from https://geologicandpaleontologicalhistory.wordpress.com/2017/04/20/faults-the-two-types/ (Accessed 28th June 2024)

cave systems with marked changes of course and many alluvial fans. Block mountains, also known as fault-block mountains, are formed by the tectonic processes acting along fault lines. The movement along these faults can cause large blocks of rock to be uplifted or to subside, resulting in the formation of ranges, such as the Vosges and Black Forest of Europe. The raised blocks are called horsts and the subsiding areas, such as the Rhine valley, are called grabens.

A further example of a faulted landscape, with a great density of faults, occurs in southwest Pakistan, where steeply dipping layers of interbedded shales and sandstones of Miocene age (the Talar Formation) have been intensely fractured and offset (Fig. 2.24). They occur in a zone called the Makran Accretionary Wedge, which is is one of the largest accretionary wedges on the globe, formed by the convergence between the Eurasian and the Arabian Plates (Back and Morley 2016).

One landscape where faulting has been a very important process is the Basin and Range Province of the western USA (Fig. 2.25). The pioneer geologist, Clarence Dutton, compared the many narrow parallel mountain ranges that distinguish the area's unique topography to an army of crawling caterpillars. In all there are more than 150 mountain ranges in the Province and many of the intemontane basins are closed and have playa lakes at their centres. As Meldahl (2011, p. 84) has remarked, 'To a degree unmatched anywhere else on Earth, the Basin and Range is a stretched land'. This area is characterized by abrupt changes in elevation, alternating between narrow faulted mountain chains (dark coloured) and flat arid valleys or basins (light coloured). The geomorphology of the province is the result of block-faulting associated with tectonic extension and lithospheric thinning that began around 17 mya in the early Miocene. The highest point fully within the province is White Mountain Peak (California) (4344 m), while the lowest point is at Badwater in Death Valley (− 86 m). Death Valley, is bounded to the west by the uplifting Panamint Range and to the east by the normal fault bounded margin of the Black Mountains. The normal fault along the Black Mountain front is highly active,

Fig. 2.24 Faults in Tertiary sedimentary rocks in southwest Pakistan. 25° 30′ 55.74″ N, 62° 28′ 51.74″ E. ©Google Earth. Scale bar = 2 km

(a)

(b)

Fig. 2.25 **a** Part of the Basin and Range Province, USA. 39° 6′ 49.37″ N, 116° 20′ 29.81″ W. ©Google Earth. Scale bar = 60 km. **b** Badwater, Death Valley, the lowest point in the Basin and Range province (at 86 m below sea level), with the Panamint Range (up to 3366 m above sea level) in the background (*Photo* A. S. Goudie)

Fig. 2.26 The Grabens, Utah. 38° 5′ 36.67″ N, 109° 54′ 40.70″ W. Scale bar = 2 km

resulting in a relatively straight fault-bounded feature from which coarse alluvial fans debouch. Normal fault scarps, where the fault plane itself is actually exposed in the rock, commonly dip between 50° and 90°, and rarely less than 40° (Blackwelder 1928). Through time, the slopes becomes eroded by transverse gullies and slope processes, and are gradually reduced in steepness, often to less than 25°. The scarps also become segmented into a series of triangular facets and the transverse valleys often assume what is called a 'wine-glass form' (Wallace 1978; Tucker et al. 2020). The facets are roughly planar surfaces with broad bases and upward-pointing apexes that occur between the valleys that drain the mountains. Excellent examples of such forms occur along the North Baikal Rift in Russia (Petit et al. 2009) and along the Wasatch Fault in Utah (Zuchiewicz and McCalpin 2000).

In the Needles District of Canyonlands National Park in Utah is a location called The Grabens (Fig. 2.26), where there is a series of normal faults that create a sort of mini Basin and Range landscape. Most of the grabens are 25 to 75 m deep and extend for about 25 km. They are associated with the presence of thick sub-surface salt beds (the Paradox Formation from the Pennsylvanian Period, 286 to 320 mya), as are the fins of rock in the nearby Moab Arches region (see Chap. 5). The Grabens began forming in the last 1.4 million years or less. They are still growing by gravity-driven extension downdip toward the Colorado River canyon. They formed and extended as the overburden spread downslope when the Colorado River cut through it into the underlying salt of the Paradox Formation (Schultz-Ela and Walsh 2002).

The most famous fault in the USA is the San Andreas Fault in California (Powell and Weldon 1992). This transcurrent fault (c 1200 km long) runs the length of the boundary between the Pacific Plate and North American Plate. It is built up of many large and small additional faults to create one big system. As the plates shift, the ground west of the San Andreas Fault moves northward. On average, movement along the fault averages about 3 to 5 cm per year. Its effects are well displayed in Tomales Bay (Fig. 2.27). During the 1906 earthquake the road across the head of Tomales Bay was offset almost 7 m. To the west of the Bay is the Pacific Plate, while to the east is the North American plate. The rock on the western shore of the Bay is Cretaceous granite, whereas on the eastern shore, the land is composed of a mix of several types of marine sedimentary rocks. The difference in relief is clear.

(a)

(b)

Fig. 2.27 **a** San Andreas Fault, Tomales Bay, California, USA. 38° 6′ 25.49″ N, 122° 47′ 17.37″ W. ©Google Earth. Scale bar = 20 km. F = fault. **b** San Andreas Fault, Bitter Creek, California, USA. 35° 6′ 29.59″ N, 119° 37′ 49.91″ W. ©Google Earth. Scale bar = 2 km

2.5 Intrusions

There are many different types of igneous masses. Some are very large (such as the 2 billion year old Bushveld Igneous Province of South Africa) and are known to have sizes up to 1600 × 160 km. The scale can decrease down to that of a volcanic plug eroded out from its surrounding ash beds. Some are intrusive and some are extrusive. The former are rocks which cool and solidify within the crust, whereas the latter are those which are poured out on to Earth's surface. It is also important to know whether the intrusions are concordant or discordant to existing structures. Major concordant intrusive forms include laccoliths, lopoliths and sills, while major discordant forms include batholiths, stocks and diapirs (intrusions in which a more mobile, ductile and deformable material is forced into brittle overlying rocks). Extrusive forms include deposits of lava, ash and ignimbrite (welded tuff). The nature of igneous rocks and igneous masses are well described by Sparks (1971) and are further discussed in Chaps. 3 and 4.

Dome mountains arise when an area of flat-lying sedimentary rocks is pushed upward by magma rising from the Earth's mantle. This magma typically fails to reach the surface and instead cools underneath, solidifying and forming the core of the mountain. Over long periods of erosion, the outer layers of the mountain are worn away, exposing the dome-shaped cooled magma of harder rock. The Black Hills of South Dakota, the Adirondacks of New York, and the La Sal Mountains of Utah are American examples of this category of mountains.

Intrusions form where molten and mobile igneous cool and solidify without breaching the ground surface to form a volcano (Fig. 2.28). The larger intrusions include batholiths and lopoliths. Batholiths (or plutons) are often granitic in composition and underlie and support many areas. When exposed at the surface by erosion they can lead to the formation of such phenomena as bornhardts and inselbergs. Lopoliths are vast, saucer-shaped and layered intrusions. Their erosion produces a series of inward-facing scarps. Small intrusions include dykes (discordant) and sills (accordant). Dykes are characteristically 1–10 m wide and often occur as swarms. Sills frequently form resistant, tabular bands within sedimentary beds. Laccoliths are sills that have thickened to form domes which arch up the overlying rocks. They are sizable magma bodies which attempted to reach the surface through a neck or conduit but spread horizontally between sedimentary layers on their way up. They pushed the host rock layers up into a domal shape before cooling beneath the surface. As a result, a 'textbook' laccolith can resemble an impact crater as the surrounding dome-shaped, upwarped eroding layers of sedimentary rock often form concentric circles around the underlying magma chamber in a way that resembles the central dome of a meteor impact (https://geology.utah.gov/map-pub/survey-notes/geosights/the-henry-mountains/) (Accessed 6th August 2024).

The Henry Mountains of Utah, USA (Fig. 2.29a), discovered by John Wesley Powell in 1869, are a famous example and were described in great detail by Gilbert (1877), who used the term laccolite rather than laccolith. Composed of porphyritic diorite, their largest peak, Mount Ellen, rises to 3512 m above sea level. They formed between 23 and 29 mya during the Oligocene when portions of subducted oceanic crust melted beneath Utah to form huge viscous masses of rising magma. A modern interpretation of the laccoliths, stocks and bysmaliths of the Henry Mountains is given by Horsman et al. (2016). Navajo Mountain in south eastern Utah is another laccolith and was intruded into and lifted up the overlying Navajo Sandstone and other sedimentary layers. The peak of Navajo Mountain, at approximately 3148 m, is comprised of uplifted Dakota Sandstone deposited during the Cretaceous Period (c 66–138 mya) (Fig. 2.29b) (https://earthobservatory.nasa.gov/images/5978/navajo-mountain-utah) (Accessed 12th July 2024).

In igneous terrains, the formation of dykes and sills has important relief implications. The Great Dyke in Zimbabwe (Wilson 1996) (Fig. 2.30) is an intrusion of igneous, metal-bearing rock that has been dated using uranium-lead isotopes to c 2.5 bya. It intrudes through the even older rocks (Archaean) of the African craton. This huge geological feature extends more than 550 km northeast to southwest across the centre of Zimbabwe, and varies from 3 to 12 km in width. It consists of a band of short, narrow ridges and hills which become taller as the range goes north, and reach up to 460 m above the Mvurwi Range. Geologically, the Great Dyke is not a dyke, but is a lopolith. The narrow linear feature in the image is actually a feeder conduit for four contiguous lopoliths intruded ito north trending fractures within the craton during an extensional tectonic phase. Subsequent deep erosion has stripped off most of the spread-out lopolithic mass, leaving only its feeder roots in the form of a 'dyke' (see Everett et al. 1986, plate T-34).

Dykes composed of dolerite are another igneous structure with relief significance, as shown by examples from Namibia and the USA. Dolerite, also known as diabase, is a fine- to medium-grained, dark gray to black intrusive igneous rock. About one-third to two-thirds of this rock type is calcium-rich plagioclase feldspar; the remainder is mostly pyroxene or hornblende. It tends to be resistant to weathering and erosion. The dolerite dykes of Namibia (Fig. 2.31) can occur in swarms as is the case with the enormous Henties Bay to Outjo swarm in the west central part of the country (Trumbull et al. 2004, 2007). This contains well

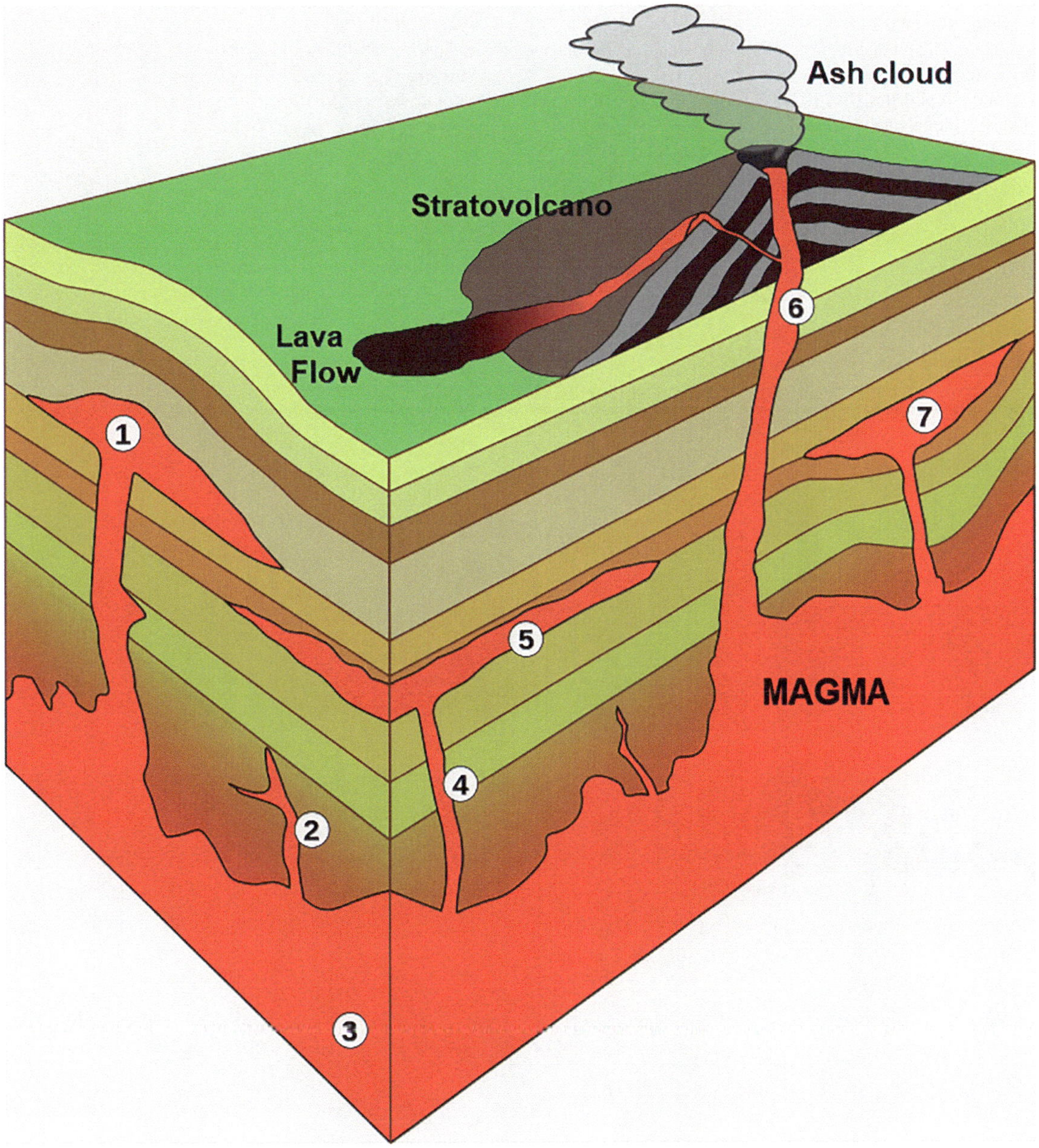

Fig. 2.28 Some of the main types of igneous intrusions. 1. Laccolith, 2. Small dyke, 3. Batholith, 4. Dyke, 5. Sill, 6. Volcanic neck, pipe, 7. Lopolith. *Source* Modified from https://en.wikipedia.org/wiki/Igneous_intrusion#/media/File:Intrusion_types.svg (Accessed 29th June 2024)

over 1000 individual dykes. They are roughly of the same age as the Etendeka Lavas and the intrusive complexes of Damaraland (see Sect. 3.4.1) and may be the exposed feeders of the Etendeka rocks. Some older dykes, of Jurassic age, occur in south eastern Namibia, where they were intruded into Dwyka and Ecca sedimentary rocks. They create hillocks and boulders.

Similarly, dykes can be seen radiating out from Ship Rock in New Mexico (Fig. 2.32). This is the remnant of an Oligocene volcano (Delaney 1987). Dykes can affect the weathering and morphology of rocks into which they are intruded, as with the Quixadá Pluton of NE Brazil (Souza et al. 2023).

One particular type of structure that may produce striking relief is the ring dyke. These are intrusive igneous bodies that are circular, oval or arcuate in plan and have steep contacts. A major circular structure in western Scotland occurs on the Ardnamurcan peninsula (Fig. 2.33). Concentric cone sheets occur around three separate centres, a map of which is given in Gale et al. (2012, Fig. 20.4). However, the precise origin of these are the subject of controversy (Magee et al. 2012; Mathieu

(a)

(b)

Fig. 2.29 **a** The northern Henry Mountains, Utah. 38° 5′ 5.94″ N, 110° 46′ 17.76″ W. ©Google Earth. Scale bar = 10 km. **b** Navajo Mountain laccolith, USA. 37° 2′ 50.55″ N, 110° 51′ 16.62″ W. ©Google Earth. Scale bar = 8 km

Fig. 2.30 Great Dyke, Zimbabwe. 18° 0′ 40.72″ S, 30° 44′ 57.42″ E. ©Google Earth. Scale bar = 100 km

et al. 2015). The volcanism that created this structure was associated with the North Atlantic Palaeogene rifting event, which also created the lava piles of Skye, Mull and Antrim.

The circular Chambe Plateau, which is about 8 km across, is part of the Mulanje Massif and is one of a number of ring dykes in southern Malawi (Fig. 2.34). This plateau is constructed of a central intrusion of syenite surrounded by three syenite ring dykes. Syenite is a coarse-grained intrusive igneous rock with a general composition similar to that of granite, but deficient in quartz. The central intrusion has been eroded down into a crater-like area. The present resistant massif of the Plateau is the result of long-term differential erosion of the syenites and the surrounding Precambrian basement gneiss (Stringer et al. 1956).

The Mesoproterozoic Pilanesberg Complex, located some 40 km NNW of Rustenburg in South Africa, is a nearly circular mass of some 25 km in diameter and rising 100–500 m from the surrounding terrain (Fig. 2.35). It originated some 1250 or 1395 mya due to an enormous volcanic eruption producing alkaline igneous rocks, including trachytes, phonolites, syenites and nepheline syenites. The dominant features of this landform complex are the concentric cone sheets formed by magma that intruded ring fractures that were created during the inward collapse of the caldera (Cawthorn 2015; Elburg and Cawthorn 2017), which gave the Pilanesberg its near perfect circular shape.

In dry regions, the sparse vegetation cover and the erosion of blankets of surficial material means that geological structures are often particularly strikingly displayed. One such structure occurs in the western part of the Sahara, where in the Adrar Region of central Mauritania there is a 40 km wide structure—Richât—that has the form of a bull's-eye target (Fig. 2.36). Also called *Guelb er Richat*, it is sometimes referred to as the 'eye of Africa'. It lies in a depression in which the drainage is annular and radial. The relief is modest—c 100 m. The area's strata, which include Ordovician sandstones, dip outwards at 10–20°, so that erosion and weathering of Late Precambrian and Early Paleozoic rocks of varying resistance has produced a series of inward-facing scarps (Sedigh et al. 2024). Quartzites form the resistant ridges. Some of the intervening annual depressions may contain seasonal lakes.

The origin of this structure is the subject of debate. Could it be volcanic, could it be an impact crater, or could it be due to differential erosion and planation of a tectonic

(a)

(b)

Fig. 2.31 **a** Dolerite dykes, Namib Desert, Namibia. 22° 30′ 45.74″ S, 14° 42′ 23.52″ E. ©Google Earth. Scale bar = 2 km. **b** Dolerite dyke near Goanikontes, Namib Desert, Namibia (*Photo* A. S. Goudie)

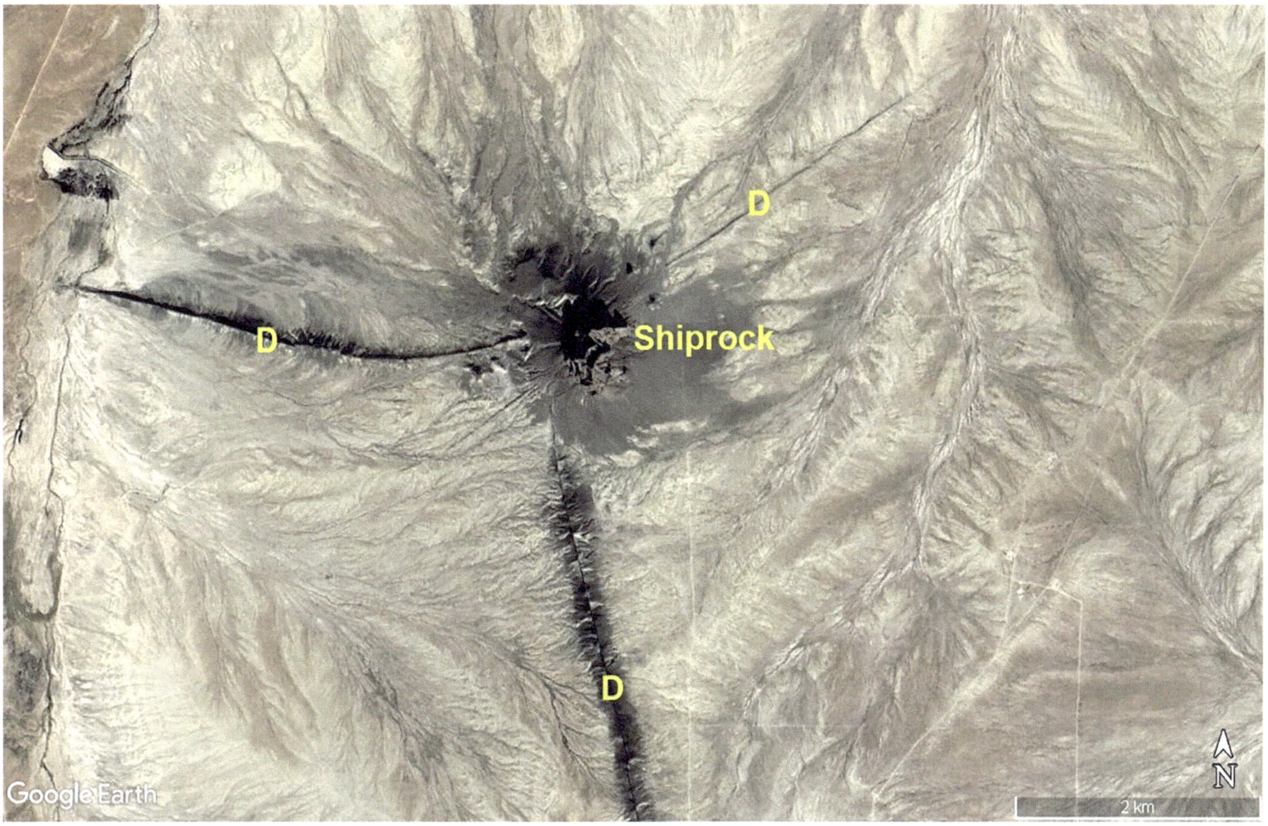

Fig. 2.32 Ship Rock, New Mexico, USA. 36° 41′ 2.70″ N, 108° 49′ 54.63″ W. ©Google Earth. Scale bar = 2 km. D = dykes

Fig. 2.33 Ardnamurcan ring dykes, Scotland. 56° 43′ 34.97″ N, 6° 6′ 53.12″ W. ©Google Earth. Scale bar = 4 km

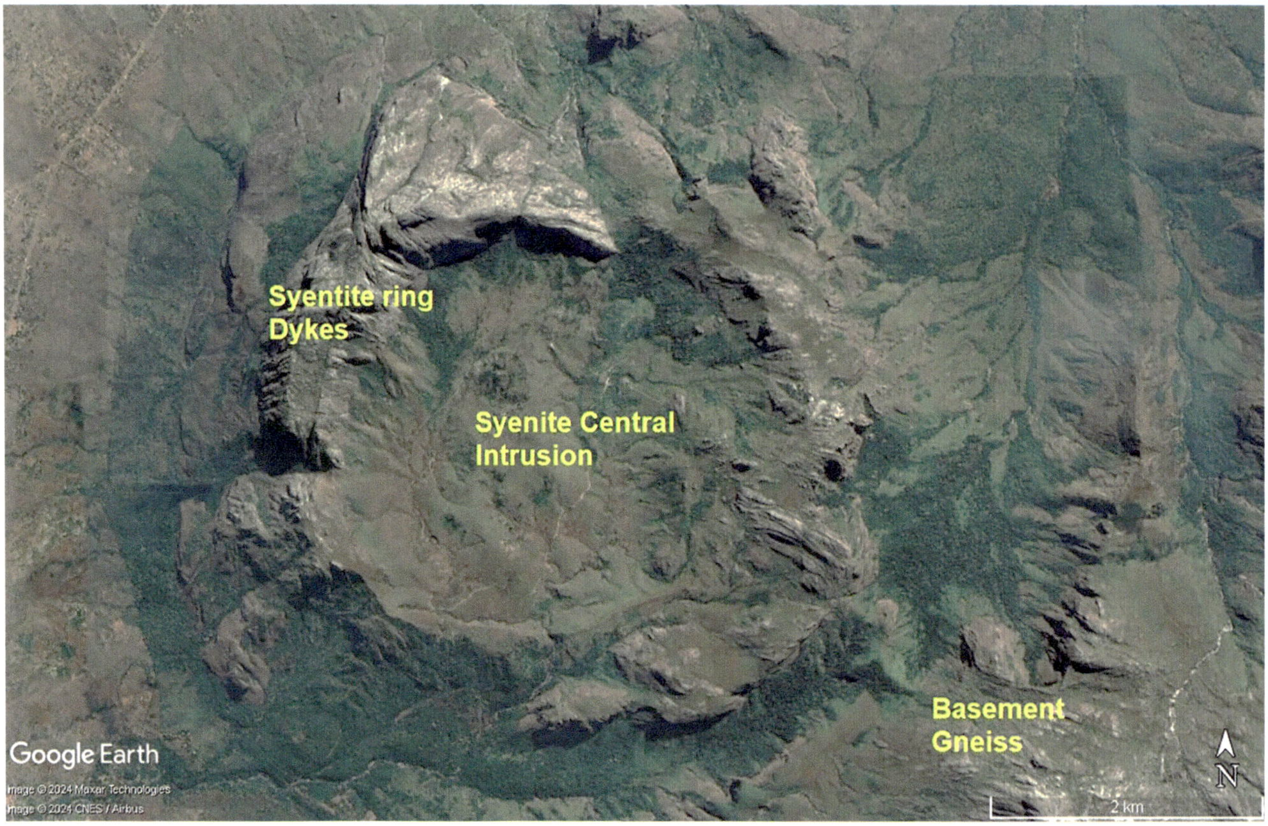

Fig. 2.34 Chambe Peak ring dyke, Malawi. 15° 54′ 38.06″ S, 35° 32′ 37.95″ E. ©Google Earth. Scale bar = 2 km

dome? The current opinion favours the third of these hypotheses, for exposed within the interior of the Richât Structure is a variety of igneous rocks. These indicate the presence of a large alkaline igneous intrusion of Cretaceous age that currently underlies the Richât Structure and created it by uplifting the overlying rock. In addition, hot fluids from the magma made their way up through the structure where they dissolved and collapsed the sedimentary rock (Matton et al. 2005; Matton and Jébrak 2014). If it had been an impact structure, the beds would surely have been more severely contorted, and there would have been masses of impact breccias.

Other ring structures are also known from the Sahara. A particularly fine group, including that at Tamgak, occurs in the Aïr Mountains of Niger (Fig. 2.37). They were formed c 450 mya (Moreau et al. 1994), and form part of a more extensive chain along the ninth meridian, extending from the north of the Aïr Massif at Adrar Bous, through southern Niger and northern Nigeria to the northwestern margin of the Benue Trough in Nigeria. Adrar Bous, a very rich area for archaeological research, consists of rocks that were intruded through the Precambrian basement into a former cover of Palaeozoic sediments. Its core consists of alkaline granites and microgranites flanked by rhyolites, dolerites, and quartziferous microgabbros to the south-west and south-east, by steeply-tilted Precambrian mica schists and amphibolites to the west, and by quartzites and basalts to the east (Williams 1971). Other intruded massifs in Aïr include Idoukal-n-Taghès (Niger's highest point at 2022 m), Tamgak (1988 m), Greboun (1944 m), Fadei, Chirriet, Taghmert, Agueraguer, Takaloukouzet, and Goundai.

At the other end of Afica, in southern Namibia, there is a c 7 km wide circular feature, Gross Brukarros, (Fig. 2.38). It contains a central depression or crater that is 3 km across and rises up about 600 m above the flatter terrain around. Unlike Roter Kamm, also in Namibia, which is a meteorite impact crater (Miller 2010), it is the result of intrusion of magma, together with superheated steam. Since the Late Cretaceous this became exposed by long term planation of the sedimentary rocks into which intrusion took place (Stengel and Busche 2002).

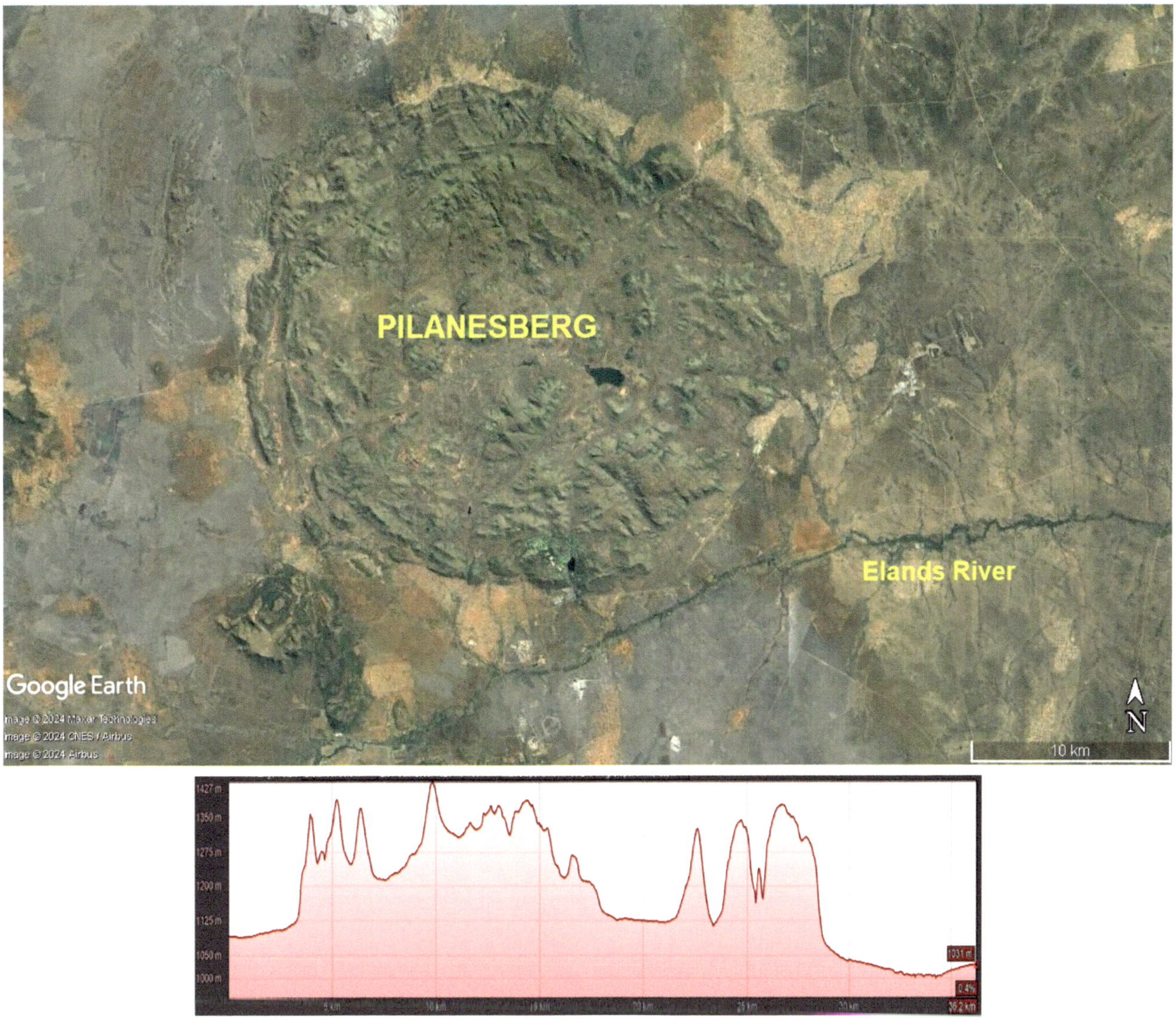

Fig. 2.35 Pilanesberg ring structure, South Africa. 25° 16′ 3.35″ S, 27° 6′ 52.37″ E. ©Google Earth. Scale bar = 10 km. Lower figure shows cross-profile from c NW to SE

Fig. 2.36 Richât structure, Mauretania. 21° 6′ 58.61″ N, 11° 22′ 0.36″ W. ©Google Earth. Scale bar = 20 km

Fig. 2.37 The Adrar Bous ring structure in Niger. 20° 20′ 51.97″ N, 9° 0′ 1.37″ E. ©Google Earth. Scale bar = 6 km

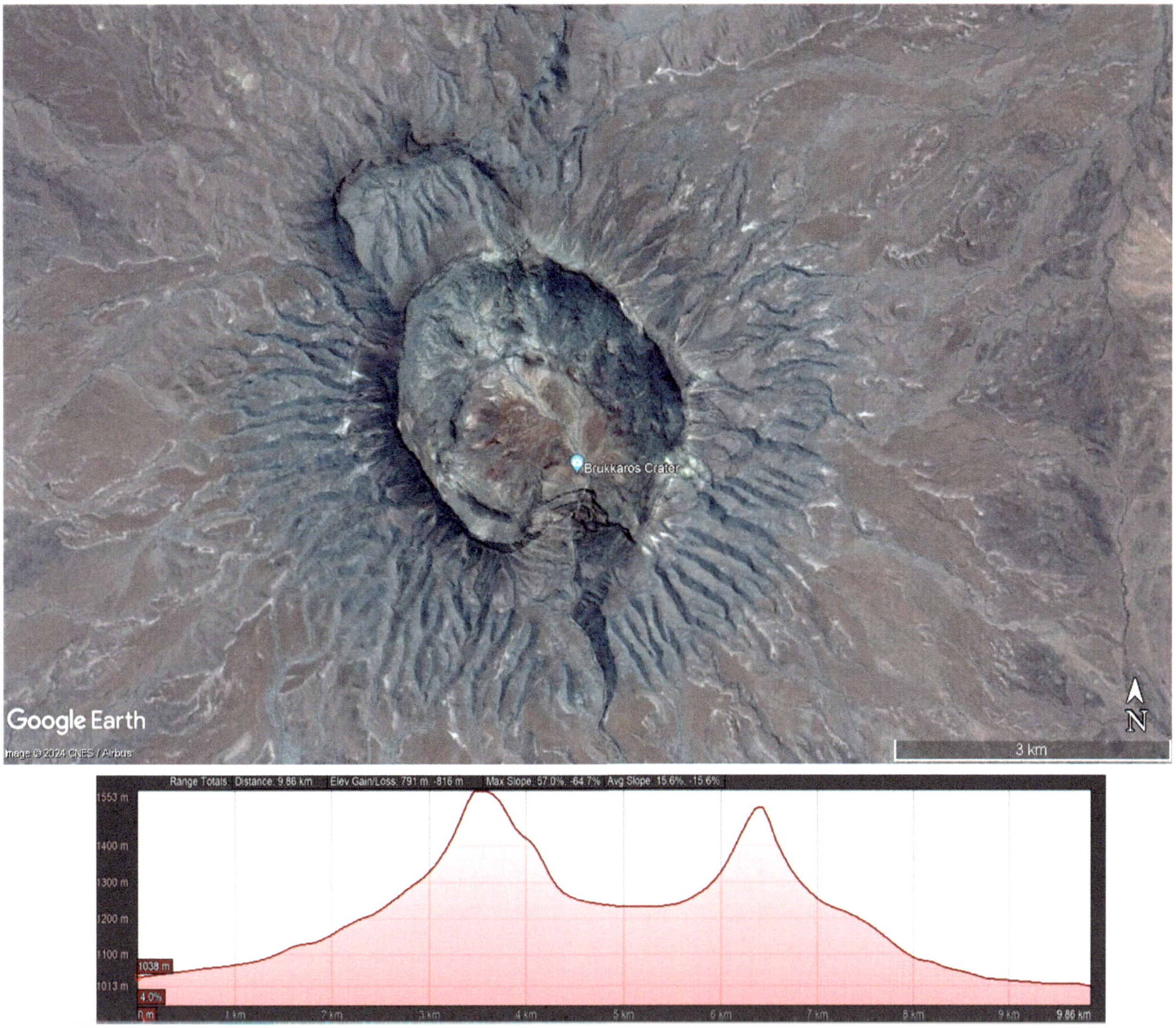

Fig. 2.38 Brukkaros, Namibia. 25° 52′ 0.54″ S, 17° 47′ 14.70″ E. ©Google Earth. Scale bar = 3 km. Lower diagram is elevation cross-profile from west to east

2.6 Salt Structures

Some other distinctive structures are associated with rocks composed of salt (halite) (Goudie 1989). For example, in southern Iran, there is a range of mountains—the Zagros—which contain beautifully displayed structures, with striking near-parallel zones of anticlines (Fig. 2.14). These mountains formed in a 'tectonic vice' as a result of the collision of the Arabian platform with the Asian landmass (Alavi 2007). This started in the Early Miocene and is still actively progressing today. The mountains are composed primarily of shelf sediments of Mesozoic and early Cenozoic age. Also in the area are some older beds, the Hormuz Formation, which consists of a sequence of evaporites that were deposited during the Late Neoproterozoic to Early Cambrian. The salt give rise to an intriguing series of landforms (Waltham 2008). Related Hormuz Formation salt domes also occur on the south side of the Straits of Hormuz in the United Arab Emirates.

The presence of these ancient evaporite layers has led to the formation of salt diapirs and salt domes. These often have clear topographic representation, especially where diapiric action has led to the piercing of the overlying strata. The structures can grow upwards very quickly, with maximum rates of 2 to 4 mm per year. When the salt reaches the surface it can be subjected to rapid solution (which produces salt karst) and to flow (to give so-called salt glaciers) The salt karst of the Zagros is well developed and includes the whole range of forms found in carbonate areas: karren, solution pipes, dolines, uvala and polje depressions, blind valleys, canyon-like forms, ponors, springs and caves (Zarei and Raeisi 2010; Bosák et al. 1999; Bruthans et al. 2024).

About 300 salt domes are identified in different places in Iran - 200 salt plugs in Zagros and the Persian Gulf region, 54 salt domes in the Great Kavir (Jackson et al. 1990), and 48 salt domes in the northwest of Iran in Zanjan and Azerbaijan Provinces (Maghsoudi 2021).The salt was laid down under marine conditions in the Eocene and Oligocene. Later Miocene sediments have been contorted by the growth and extrusion of the salt domes (Fig. 2.39a).

The growth of salt structures can create characteristic drainage patterns and slope forms. A model of this was proposed by Berger and Aghassy (1984) who envisaged three development phases. First, in the 'positive relief stage' there is a central dome on which radial drainage by outbound consequent streams is dominant, and these dissect long isoclinal slopes. Second, in the 'breached stage' the initial topographic high in the centre of the dome becomes lowered, an inversion of topography begins to occur and a major depression develops at the dome's centre. Inward-facing scarp slopes and inbound obsequent streams develop. Third, in the 'obliterative stage' these inbound obsequents expand headwards and capture much of the consequent outward-bound drainage. Sedimentation, flood-plain development and marsh formation occur in the core.

If the rate of diapiric growth is faster than the rate of salt solution, salt will be extruded at the ground surface. This may tend to be favoured in arid areas, such as central Iran, where low rainfall levels cause low rates of dissolution. Because of its mechanical properties, such extruded salt may begin to flow, creating 'salt (Fig. 2.39 b). These are termed 'namakiers' (from *namak*—the Farsi word for salt—and glacier) (Talbot 1979). Some of these in the Zagros Mountains' are huge. The example at Kuh-e-Namak is 2000 m long, 3500 m wide and up to 50 m thick, but perhaps the biggest example is Kuh-e-Gach which is also 50 m thick, but attains a width of around 4700 m and has an area of around 23.5 km^2. Their velocity of movement is less than that of true glaciers, the average yearly speed being only a few metres with movement tending to occur after rainfall events. However, they have some similarities to real glaciers. They display complex folds where they flow over bedrock irregularities, and at their distal ends they may feather out in a mass of unbedded detritus analogous to a terminal moraine. Salt glaciers are also subject to retreat if the balance of wastage by solution should exceed that of outward flow of salt from the diapir, and this explains the presence of isolated exotic blocks, analogous to glacial erratics, some kilometers beyond the plugs themselves.

Other areas with numerous salt piercement structures occur in the southern USA. One of these is the Avery Salt Mountain, the home of Tabasco Sauce (Fig. 2.40). This is one of The Five Islands salt domes that are located near the western boundary of the Mississippi River delta plain in coastal south Louisiana The islands, aligned in an approximately NW–SE trend, are named Jefferson Island, Avery Island, Weeks Island, Cote Blanche Island, and Belle Isle (Autin 2002).

Other notable domes occur, *inter alia*, in Melville Island, Arctic Canada (Fig. 2.41), and at Rocher de Sel (Salt Rock) in Algeria. The former (https://earthobservatory.nasa.gov/images/6871/salt-domes-on-melville-island) (Accessed 2nd May 2024) lies in the Sverdrup Basin, a southwest-north-east-trending basin about 1300 km long and 400 km wide. Its salt deposits are of Carboniferous age (359 to 299 mya). The Rocher de Sel, known in Arabic as Khanguet-el-Melah ('Parade of Salt'), lies in the Ouled Naïl Range, near the town of Djelfa. It is nearly 1500 m in diameter and 105 m tall (Gautier 1914). It has a high density of sinkholes across its surface and a series of springs and resurgences near the contact between the salt and the surrounding claystone rock (Fig. 2.42).

Indeed, as the Rocher de Sel showed, in areas where highly soluble thick halite (salt) or gypsum deposits have accumulated, solution may cause sinkholes to develop. An example of this is provided by the Dead Sea basin, where the recent decline in level caused by water abstraction has promoted their formation. Yizhaq et al. (2017) reported the presence of 5,500 sinkholes along the Dead Sea in Israel/Palestine, and that since 2013, about 400 new sinkholes have appeared each year causing damage to infrastructure, palm plantations, and tourist facilities. As Yechieli et al. (2016, p. 1) explain it in the context of the Israeli/Palestinian side of the basin:

> The eastward retreat of the shoreline and the lake-level drop (1 m/year in recent years) cause an eastward and downward migration of the fresh/saline groundwater interface. Consequently, a subsurface salt layer, which was previously enveloped by saline groundwater, is gradually being invaded and submerged by relatively fresh groundwater, and cavities form due to the rapid dissolution of the salt. Collapse of the overlying sediments into these cavities results in sinkholes at the surface.

Likewise, Watson et al. (2019) mapped the formation of c 1,150 sinkholes and five uvalas that formed from c 1980 to 2017 on the Jordanian shore (Fig. 2.43). Full details of the distribution of the sinkholes and their morphometric characteristics are given by Sevil and Gutiérrez (2023).

2.7 Conclusion

This chapter has demonstrated, through examples from all over the world, that geological structures play a fundamental role in controlling the effects of rocks on relief. It has drawn attention to the significance of joints, folds (including anticlines and synclines), faults, and intrusive domes (including those composed of salt), in determining the role that rocks play in landscape development.

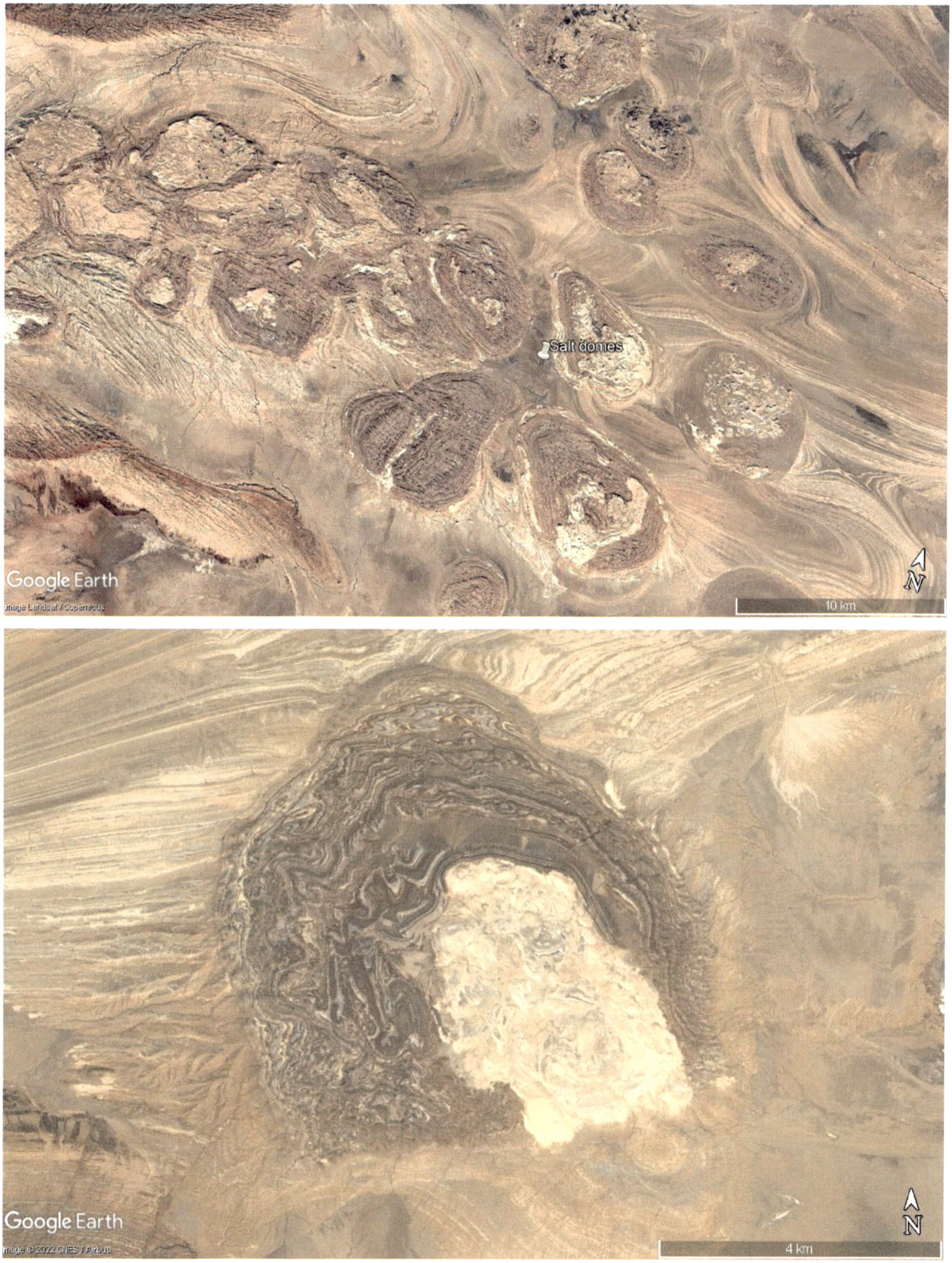

Fig. 2.39 **a** Iran salt domes. 34° 56′ 41.11″ N, 53° 39′ 23.83″ E. ©Google Earth. Scale bar = 10 km. **b** Iran salt dome with salt glacier on south east side. 34° 58′ 0.39″ N, 54° 3′ 31.86″ E. ©Google Earth. Scale bar = 4 km

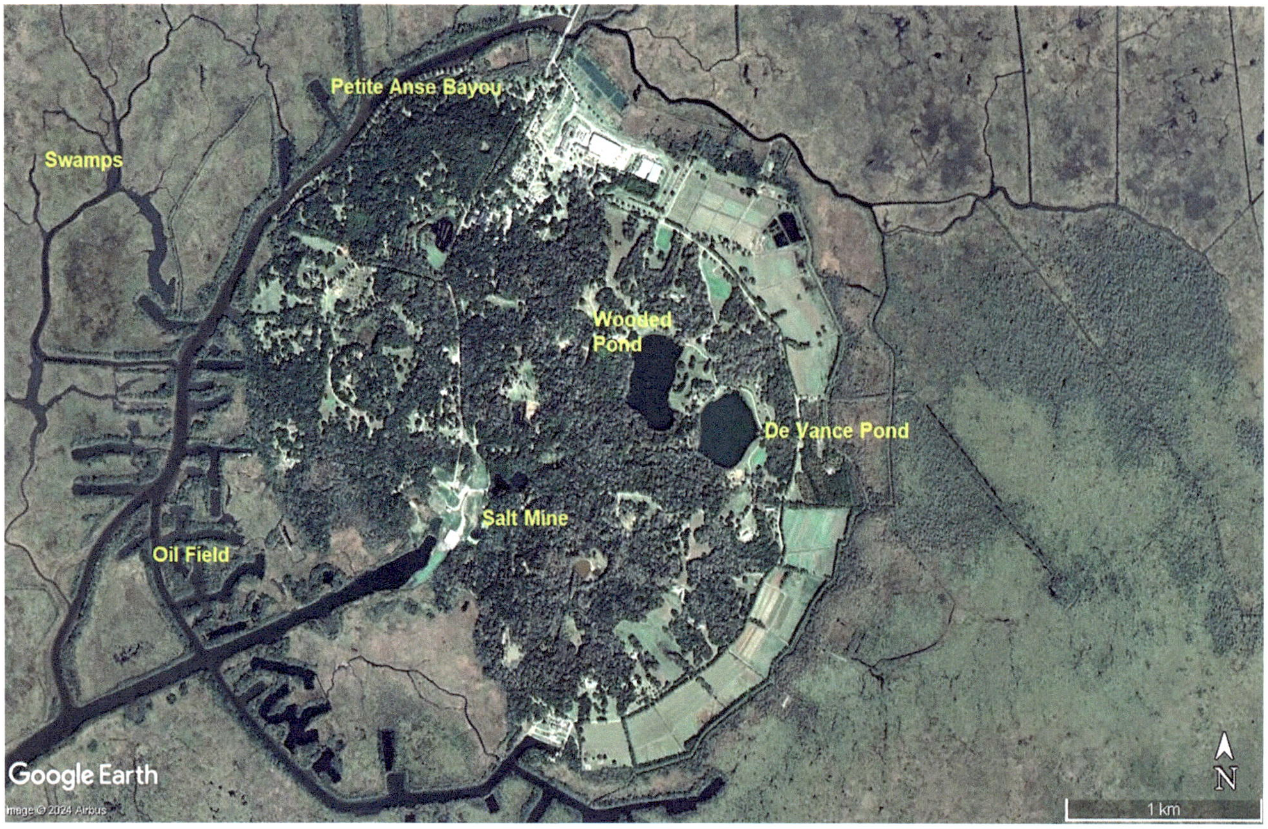

Fig. 2.40 Avery Island salt dome, USA. 29° 53′ 53.11″ N, 91°54′ 2.46″ W. ©Google Earth. Scale bar = 1 km

Fig. 2.41 Melville Island salt domes, northern Canada. 76° 42′ 34.96″ N, 108° 47′ 22.24″ W. ©Google Earth. Scale bar = 10 km

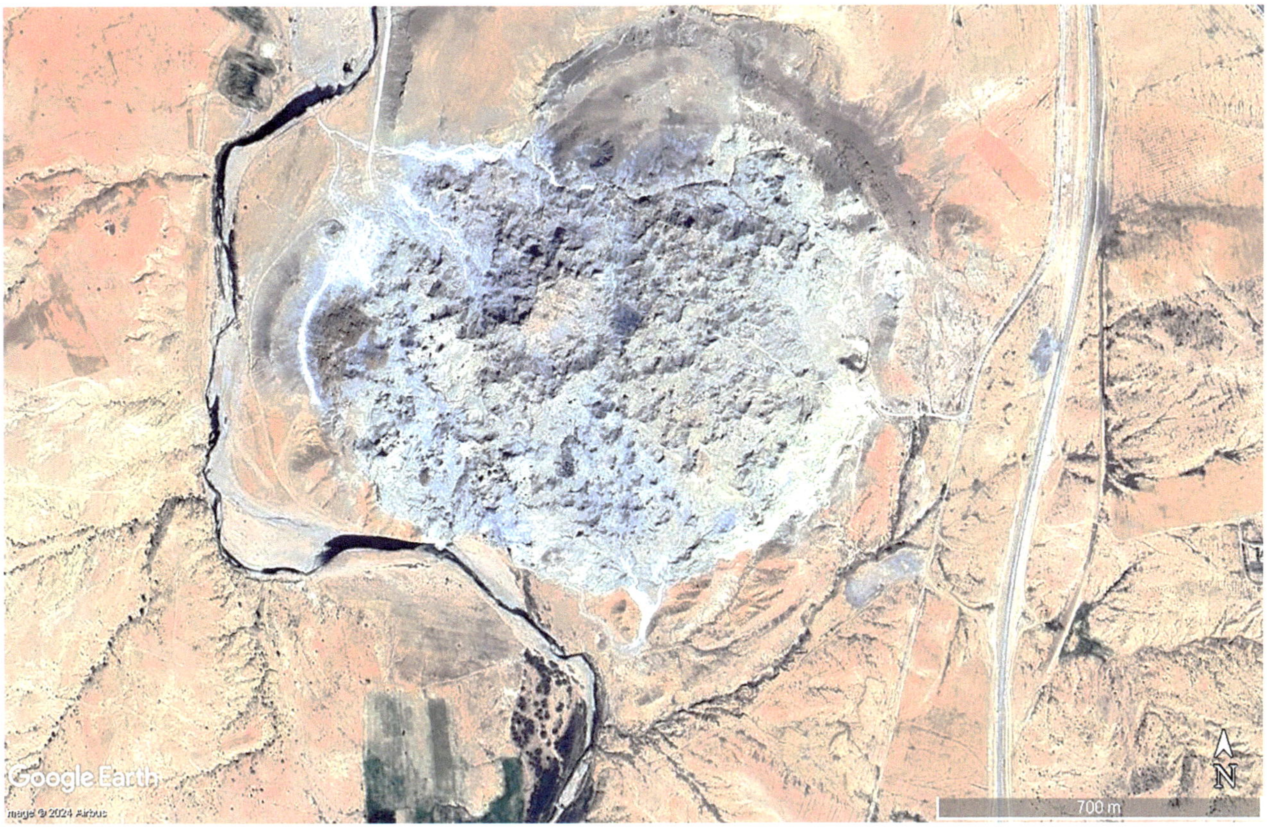

Fig. 2.42 Rocher de Sel salt dome, Algeria. 34° 49′ 59.60″ N, 3° 5′ 39.58″ E. ©Google Earth. Scale bar = 700 m

Fig. 2.43 Sinkholes (dolines) and uvalas formed by the ongoing desiccation of the Dead Sea, Jordan. 31° 20′ 55.65″ N, 35° 32′ 39.62″ E. ©Google Earth. Scale bar = 200 m

References

Alavi M (2007) Structures of the Zagros fold-thrust belt in Iran. Am J Sci 307(9):1064–1095

Anhaeusser CR (2024) Satellite imagery reveals the nature of the Archaean granite-greenstone basement of the Kruger National Park, South Africa, including the newly discovered Olifants River Greenstone Belt. S Afr J Geol. https://doi.org/10.25131/sajg.127.0008

Autin WJ (2002) Landscape evolution of the Five Islands of south Louisiana: scientific policy and salt dome utilization and management. Geomorphology 47(2–4):227–244

Back S, Morley CK (2016) Growth faults above shale—seismic-scale outcrop analogues from the Makran foreland, SW Pakistan. Marine Petrol Geol 70:144–162

Baig SS, Xue C, Alam M, Ullah N, Alam M, Hussain SA (2020) Evolution of Karakoram sequence, Hunza Valley NE Pakistan: evidences from field mapping and micro-analytical work. J Mountain Area Res 5:19–27

Berger Z, Aghassy J (1984) Near surface moisture and evolution of structurally controlled drainage in soft sediments. In: LaFleur RG (ed) Groundwater as a geomorphic agent. Allen and Unwin, London, pp 59–77

Blackwelder E (1928) The recognition of fault scarps. J Geol 36(4):289–311

Bosák P, Bruthans J, Filippi M, Svoboda T, Šmíd J (1999) Karst and caves in salt diapirs, SE Zagros Mts. (Iran). Acta Carsologica 28(2). https://doi.org/10.3986/ac.v28i2.484

Brunsden D, Coombe K, Goudie AS, Parker AG (1996) The structural geomorphology of the Isle of Portland, southern England. Proc Geol Ass 107(3):209–230

Bruthans J, Filippi M, Slavík M, Závada P, Zare M (2024) Rapid evolution of salt glacier caves on a mountain diapir in a semiarid climate. Geomorphology 448:109058

Carr MJ (2017) A Google Earth database of Central American volcanic vents. Rev Geol Am Central 56:7–15

Cawthorn RG (2015) The geometry and emplacement of the Pilanesberg Complex, South Africa. Geol Mag 152(5):802–812

Cody E, Draebing D, McColl S, Cook S, Brideau MA (2020) Geomorphology and geological controls of an active paraglacial rockslide in the New Zealand Southern Alps. Landslides 17:755–776

Delaney PT (1987) Ship Rock, New Mexico: the vent of a violent volcanic eruption. In: Rocky Mountain Section of the Geological Society of America. Geological Society of America, pp 411–415

Duszyński F, Migoń P (2020) Sandstone landforms of the High Weald. In: Goudie AS, Migoń P (eds) Landscapes and landforms of England and Wales. Springer, Cham, pp 103–118

Duszyński F, Migoń P, Strzelecki MC (2019) Escarpment retreat in sedimentary tablelands and cuesta landscapes–landforms, mechanisms and patterns. Earth Sci Rev 196:102890

Elburg MA, Cawthorn RG (2017) Source and evolution of the alkaline Pilanesberg Complex, South Africa. Chem Geol 455:148–165

Evans IS (1994) Lithological and structural effects on forms of glacial erosion: cirques and lake basins. In: Robinson DA, Williams RBG (eds) Rock weathering and landform evolution. John Wiley and Sons, Chichester, pp 455–472

Everett JR, Morisawa M, Short NM (1986) Tectonic landforms. In: Short NM, Blair RW (eds) Geomorphology from space. A global overview of regional landforms. NASA, Washington DC, p 717

Finzi Y, Harlev N (2016) A regional approach for modeling cliff retreat rate: the Makhteshim Country, Israel. Geomorphology 271:65–73

Gale AS, Anderton R, Strachan RA (2012) Palaeogene and Neogene events: the North Atlantic plume and Alpine pulses. In: Woodcock N, Strachan R (eds) Geological history of Britain and Ireland. Blackwell, Oxford, pp 387–408

Gale SJ, Drysdale RN, Scherrer NC, Fischer MJ (1997) The lost city of north-west Queensland: a test of the model of giant grikeland development in semi-arid Karst. Austral Geogr 28(1):107–115

Gautier EF (1914) Le rocher de sel de Djelfa. Ann Géogr 23:245–260

Gilbert GK (1877) Geology of the Henry Mountains. Government Printing Office, Washington DC, p 160

Glasser NF, Ghiglione MC (2009) Structural, tectonic and glaciological controls on the evolution of fjord landscapes. Geomorphology 105(3–4):291–302

Goudie AS (1989) Salt tectonics and geomorphology. Progr Phys Geogr 13(4):597–605

Goudie AS et al (1984) The geomorphology of the Hunza Valley, Karakoram Mountains, Pakistan. In: Miller KJ (ed) The international Karakoram project, vol 2. Cambridge Univ Press, Cambridge, pp 358–410

Gregory JW (1913) The nature and origin of fiords. Murray, London, p 542

Hawley CC, Robeck RC, Dyer HB (1968) Geology, altered rocks and ore deposits of the San Rafael Swell, Emery County, Utah. US Geol Surv Bull 1239:166

Hooyer TS, Cohen D, Iverson NR (2012) Control of glacial quarrying by bedrock joints. Geomorphology 153:91–101

Horsman E, Broda RJ, Gwyn NZ, Maurer EA, Thornton ED, Ward MT (2016) Progressive construction of laccolithic intrusive centers: Henry Mountains, Utah, U.S.A. In: Breitkreuz C, Rocchi S (eds) Physical geology of shallow magmatic systems. Springer, Cham, pp 327–347

Jackson MPA, Cornelius RR, Craig CH, Gansser A, Stöcklin J, Talbot CJ (1990) Salt diapirs of the Great Kavir, central Iran. Geol Soc Am Mem 177:139

Jansen JD et al (2022) Geomorphic imprints of lithospheric flexure in central Australia. Earth Planet Sci Lett 584:117456

Jones DKC (2020) The Weald. In: Goudie AS, Migoń P (eds) Landscapes and landforms of England and Wales. Springer, Cham, pp 73–101

Krabbendam M, Glasser NF (2011) Glacial erosion and bedrock properties in NW Scotland: abrasion and plucking, hardness and joint spacing. Geomorphology 130(3–4):374–383

Krasnov BR, Mazor E (eds) (2001) The Makhteshim country: a laboratory of nature: geological and ecological studies in the desert region of Israel. Pensoft, Sofia, p 411

Lageson DR, Larsen MC, Lynn HB, Treadway WA (2012) Applications of Google Earth Pro to fracture and fault studies of Laramide anticlines in the Rocky Mountain foreland. Geol Soc Am Special Paper 492:209–220

Lawal MA et al (2022) A simplified GIS and Google Earth-based approach for lineaments and terrain attributes mapping in a basement complex terrain. Sci Reports 12(1):15801

Li A, Ran Y, Gomez F, Thompson Jobe JA, Liu H, Xu L (2020) Segmentation of the Kepingtage thrust fault based on paleoearthquake ruptures, southwestern Tianshan, China. Nat Hazards 103:1385–1406

Magee C, Stevenson C, O'Driscoll B, Schofield N, McDermott K (2012) An alternative emplacement model for the classic Ardnamurchan cone sheet swarm, NW Scotland, involving lateral magma supply via regional dykes. J Struct Geol 43:73–91

Maghsoudi M (2021) Other desert landforms. In: Maghsoudi M (ed) Desert landscapes and landforms of Iran. Springer, Cham, pp 133–145

Manjoro M (2015) Structural control of fluvial drainage in the western domain of the Cape Fold Belt, South Africa. J Afr Earth Sci 101:350–359

Mathieu L, Burchardt S, Troll VR, Krumbholz M, Delcamp A (2015) Geological constraints on the dynamic emplacement of cone-sheets—the Ardnamurchan cone-sheet swarm, NW Scotland. J Struct Geol 80:133–141

Matton G, Jébrak M (2014) The "eye of Africa" (Richat dome, Mauritania): an isolated Cretaceous alkaline–hydrothermal complex. J Afr Earth Sci 97:109–124
Matton G, Jébrak M, Lee JK (2005) Resolving the Richat enigma: doming and hydrothermal karstification above an alkaline complex. Geology 33(8):665–668
Meldahl KH (2011) Rough hewn land. A geological journey from California to the Rocky Mountains. University of California Press, Berkeley, p 296
McG MR (2010) Roter Kamm impact crater of Namibia. New data on rim structure, target rock geochemistry, ejecta and meteorite trajectory. Geol Soc Am Spec Paper 465:489–508
Moon BP, Selby MJ (1983) Rock mass strength and scarp forms in southern Africa. Geogr Ann Series A 65(1–2):135–145
Moreau C, Demaiffe D, Bellion Y, Boullier AM (1994) A tectonic model for the location of Palaeozoic ring complexes in Aïr (Niger, West Africa). Tectonophysics 234(1–2):129–146
Morisawa M (1989) Rivers and valleys of Pennsylvania, revisited. Geomorphology 2:1–22
Muir-Wood R (2024) This volcanic isle. Oxford University Press, Oxford, p 338
Nicholas RM, Dixon JC (1986) Sandstone scarp form and retreat in the land of standing rocks, Canyonlands National Park,Utah. Z Geomorph 30(2):167–187
Nilsen TH (1973) The relation of joint patterns to the formation of fjords in western Norway. Norsk Geol Tids 53(2):183–194
Petit C et al (2009) Height of faceted spurs, a proxy for determining long-term throw rates on normal faults: evidence from the North Baikal Rift System, Siberia. Tectonics 28(6). https://doi.org/10.1029/2009TC002555
Pickup G, Allan G, Baker VR (1988) History, palaeochannels and palaeofloods of the Finke River, central Australia. In: Warner RF (ed) Fluvial geomorphology of Australia. Academic Press, Sydney, pp 177–200
Powell RE, Weldon RJ (1992) Evolution of the San Andreas Fault. Ann Rev Earth Planet Sci 20:431–468
Quick LJ, Eckardt FD (2015) The Cederberg: a rugged sandstone topography. In: Grab S, Knight J (eds) Landscapes and landforms of South Africa. Springer, Cham, pp 85–93
Rollinson H (2014) A (virtual) field excursion through the Oman ophiolite. Geol Today 30(3):110–118
Samal AK, Srivastava RK, Ernst RE, Söderlund U (2019) Neoarchean-Mesoproterozoic mafic dyke swarms of the Indian Shield mapped using Google Earth™ images and ArcGIS™, and links with large igneous provinces. In: Srivastava RK, Ernst RE, Peng P (eds) Dyke swarms of the world: a modern perspective. Springer, Singapore, pp 335–390
Scharf TE, Codilean AT, De Wit M, Jansen JD, Kubik PW (2013) Strong rocks sustain ancient postorogenic topography in southern Africa. Geology 41(3):331–334
Schultz-Ela DD, Walsh P (2002) Modeling of grabens extending above evaporites in Canyonlands National Park, Utah. J Struct Geol 24:247–275
Scott AW (2015) Kimberley coast. Envirobook, Sussex Inlet, New South Wales, p 124
Searle MJ (1991) Geology and tectonics of the Karakoram Mountains. Wiley, Chichester, p 358
Searle M (2013) Colliding continents. Oxford University Press, Oxford, p 438
Searle M, Cox J (1999) Tectonic setting, origin, and obduction of the Oman ophiolite. Geol Soc Am Bull 111(1):104–122
Sedigh ME, Ouali H, Jébrak M, Mokhtari A (2024) Bulk composition and O-C isotope constraints on the petrogenesis of Richat carbonatites (Mauritania). J Afr Earth Sci 209:105126
Selby MJ (1982) Rock mass strength and the form of some inselbergs in the central Namib Desert. Earth Surf Proc Landf 7(5):489–497
Sevil J, Gutiérrez F (2023) Morphometry and evolution of sinkholes on the western shore of the Dead Sea. Implications for susceptibility assessment. Geomorphology 434:108732
Souza AS, Maia RP, Bezerra FH, Migoń P, Siame LL (2023) Granitic inselberg erosion controlled by dike swarm array in semiarid Brazil. Geomorphology 440:108865
Sparks BW (1971) Rocks and relief. Longman, London, p 422
Stengel I, Busche D (2002) Namibia's oldest inselbergs—exhumed landforms of a Late Proterozoic etchplain. In: Yang X (ed) Desert and alpine environments. China Ocean Press, Beijing, pp 121–143
Stringer KV, Holt DN, Groves AW (1956) The Chambe plateau ring complex of Nyasaland. Colonial Geol Min Res 6:3–18
Summerfield MA (1991) Global geomorphology. Longman, Harlow, p 537
Talbot CJ (1979) Fold trains in a glacier of salt in southern Iran. J Struct Geol 1:5–18
Thornbury WD (1965) Regional geomorphology of the United States. Wiley, Chichester, p 610
Trumbull RB, Vietor T, Hahne K, Wackerle R, Ledru P (2004) Aeromagnetic mapping and reconnaissance geochemistry of the Early Cretaceous Henties Bay-Outjo dike swarm, Etendeka igneous province, Namibia. J Afr Earth Sci 40(1–2):17–29
Trumbull RB, Reid DL, De Beer C, Van Acken D, Romer RL (2007) Magmatism and continental breakup at the west margin of southern Africa: a geochemical comparison of dolerite dikes from northwestern Namibia and the Western Cape. S Afr J Geol 110(2–3):477–502
Tucker GE, Hobley DE, McCoy SW, Struble WT (2020) Modeling the shape and evolution of normal-fault facets. J Geophys Res: Earth Surf 125(3):e2019JF005305
van Rooyen D, Sappin AA, Corrigan D, Rayner N, Houlé MG (2024) The corrugated hills continental flood basalt: a c. 2.17 Ga flood basalt province related to breakup of the Superior craton. Geol Soc Lond Spec Pub 542(1):665–699
Wallace RE (1978) Geometry and rates of change of fault-generated range fronts, north-central Nevada. J Res US Geol Surv 6(5):637–650
Waltham T (2008) Salt terrains of Iran. Geol Today 24(5):188–194
Watson RA et al (2019) Sinkholes and uvalas in evaporite karst: spatio-temporal development with links to base-level fall on the eastern shore of the Dead Sea. Solid Earth 10(4):1451–1468
Williams MA (1971) Geomorphology and quaternary geology of Adrar Bous. Geogr J 137:450–455
Wilson AH (1996) The Great Dyke of Zimbabwe. Dev Petrol 15:365–402
Woor S, Thomas DS, Parton A, Leenman A (2023) Morphology and controls of the mountain-front fan systems of the Hajar Mountains, south-east Arabia. Earth Sci Rev 237:104316
Yechieli Y, Abelson M, Baer G (2016) Sinkhole formation and subsidence along the Dead Sea coast, Israel. Hydrogeol J 24(3):601
Yizhaq H, Ish-Shalom C, Raz E, Ashkenazy Y (2017) Scale-free distribution of Dead Sea sinkholes: observations and modeling. Geophys Res Lett 44(10):4944–4952
Zarei M, Raeisi E (2010) Karst development and hydrogeology of Konarsiah salt diapir, south of Iran. Carb Evap 25:217–229
Zilberman E (2024) "Makhteshim"—unique arid land erosion cirques in the Negev. In: Frumkin A, Shtober-Zisu N (eds) Landscapes and landforms of Israel. Springer, Cham, pp 297–311
Zuchiewicz WA, McCalpin JP (2000) Geometry of faceted spurs on an active normal fault: case study of the Central Wasatch Fault, Utah, USA. Ann Soc Geol Poloniae 70:231–249

Igneous and Metamorphic Rocks

3

Abstract

This chapter is primarily concerned with intrusive igneous rocks and the landforms to which they give rise, including tors, inselbergs, bornhardts, and the like. However, it also considers the nature of the landforms associated with various metamorphic rocks, including schists and gneisses. Particular attention is given to inselbergs in the Namib Desert, Zimbabwe, Morocco, and parts of the USA and to the various weathering forms that develop on igenous rocks.

Keywords

Igneous · Metamorphic · Granite · Migmatite · Inselberg · Tor · Bornhardt · Schist · Gneiss

3.1 Introduction

Igneous and metamorphic rocks can be associated with some distinctive landforms. One has only to think, for example, of the tors which are such a feature of the granites of Dartmoor (Gunnell and Jarman 2020) (Fig. 3.1) and Bodmin Moor (Evans 2020) in south west England.

3.2 Igneous Rocks

Igneous rocks are those formed from crystallization of magma into rock-forming minerals. Extrusive rocks are those which are poured out from the earth's surface (e.g. at volcanoes—see Chap. 4), whereas intrusive igneous rocks are those which cool and solidify within the crust (e.g. granites). The former may affect relief immediately, the latter only secondarily after they have been exposed by denudation. Intrusive igneous bodies, called plutons, may vary hugely in size, and such large bodies include batholiths (see Sparks 1971, Chap. 3). A batholith is an exposed area of (mostly) continuous plutonic rock that covers an area larger than 100 km^2. Areas smaller than 100 km^2 are called stocks. The largest batholiths may extend over hundreds of km. The coastal batholith of Peru is c 500 km long and c 20–30 km wide. Another mammoth batholith is the Sierra Nevada Batholith, which is a continuous granitic formation that makes up much of the Sierra Nevada in California. An even larger batholith is found predominantly in the Coast Mountains of western Canada; it extends for 1800 km and reaches into south eastern Alaska.

Granites are relatively coarse-grained intrusive igneous rocks composed of quartz (20 to 60%) and feldspars. Granites have been sub-divided into two types defined by the percentage of plagioclase among total feldspar, namely syenogranite (10–35%) and monzogranite (35–65%). Granodiorite differs from granite through the higher percentage of plagioclase in respect to potassium feldspar. Granites may be highly jointed or may be massive in form. They may display sheet fractures which are arcuate fractures defining slabs of rock that range from 0.5 to 10 m thick. They normally form in sets parallel to the Earth's surface but may form in convex-upward or concave-upward sets. The most popular hypothesis for their development is that they are the result of expansion and tangential fracturing consequent on erosional offloading (pressure release). Exfoliation may develop and this is an important issue with respect to their geomorphological effects (see Fig. 3.17). Full discussions of granites and their landform impacts are given in Migoń (2006) and Twidale and Romaní (2005).

3.3 Metamorphic Rocks

Metamorphic rocks are rocks that have undergone a profound transformation in mineral composition, texture, and sometimes even chemical structure without melting. This

A. Goudie, *Rocky Landscapes of the World with Google Earth*, https://doi.org/10.1007/978-3-031-77418-8_3

Fig. 3.1 Haytor, Dartmoor, Devon, England. 50° 34′ 47.07″ N, 3° 45′ 19.64″ W. ©Google Earth. Scale bar = 100 m

transformation occurs in response to changes in geological conditions, primarily elevated temperature and pressure (https://geologyscience.com/rocks/metamorphic-rocks/schist/?amp) (Accessed 18th June 2024). Metamorphism typically affects pre-existing rocks, which can be of sedimentary, igneous, or metamorphic origin, and results in the formation of new minerals and textures. The original rock from which a metamorphic rock forms is called the protolith. Contact metamorphism occurs when rocks are subjected to high temperatures due to their proximity to molten magma or lava. The heat from the molten material causes the surrounding rocks to undergo metamorphism without a significant increase in pressure. Regional metamorphism is the most widespread type of metamorphism and occurs over large areas due to tectonic forces associated with mountain-building events and the collision of tectonic plates. It involves both high pressure and temperature. Dynamic metamorphism occurs when rocks are subjected to extreme pressure without a significant increase in temperature. This pressure is typically associated with fault and shear zones, where rocks are deformed and crushed. Hydrothermal metamorphism involves the alteration of rocks by hot, chemically active fluids, typically groundwater or hydrothermal solutions rich in dissolved minerals. These fluids can react with the surrounding rock, changing its mineral composition. Burial metamorphism occurs as rocks are buried deep within the Earth's crust due to deposition of sediments or to subsidence. The increased pressure and temperature at depth can lead to mineral changes. Shock metamorphism is a rare type of metamorphism that occurs when rocks are subjected to the extreme pressures and temperatures associated with meteorite impacts.

Schist is one type of metamorphic rock. It is characterized by its foliated texture, which means it possesses distinct layers or bands of minerals that have undergone significant physical and chemical changes due to heat, pressure, and other geological processes. The term 'schist' is derived from the German word 'schizein,' meaning 'to split,' and refers to the rock's tendency to break easily along its foliation planes. Intense heat and pressure cause the minerals within the rock to re-crystallize and align themselves in parallel layers, giving schist its characteristic foliation. The minerals that make up schist can vary widely, but common ones include mica (such as biotite and muscovite), quartz, and feldspar.

Slate is a fine-grained, foliated, homogeneous, metamorphic rock derived from an original shale-type sedimentary rock composed of clay or volcanic ash through low-grade, regional metamorphism. The foliation in slate, called 'slaty cleavage', is caused by strong compression in which fine-grained clay forms flakes to regrow in planes perpendicular to the compression. The cleavage in slates can make them ustable and can lead to toppling, translational sliding, sackung formation, and the development of uphill-facing scarps, as has been found in Japan (Yokoyama 2020), Taiwan (Lo and Feng 2014) and Spain (Calcaterra et al. 1998). Slates may also be rendered unstable by freeze–thaw and other weathering processes, and this also leads to landsliding, as found in Gansu, China (Cui et al. 2017). Related to slates are phyllites. These are a type of foliated metamorphic rock formed from slate that is further metamorphosed so that very fine-grained white mica achieves a preferred orientation. They represent a gradation in the degree of metamorphism between slate and schist. They also appear to be very significant in terms of landslide distribution in areas such as the Himalayas (Gerrard 1994) and western China (Guan et al. 2023).

Another important metamorphic rock is gneiss. Gneisses are characterised by compositional layering produced either by deformation of a pre-existing texture or structure, or by metamorphic segregation, or by a combination of both. They tend to be coarser grained than schists. Gneiss domes, sometimes enormous, are common in orogenic belts (Whitney et al. 2004). As Teyssier and Whitney (2002) noted, gneiss domes occur worldwide in orogens ranging in age from Archean through Cenozoic and in tectonic settings from wide (hundreds to thousands of km) collisional orogens to narrow (tens of km) shear belts. Well-documented examples occur in the North American Cordillera, Himalayas, Pamirs, Alps, Iberian and French Variscides, the Bering Sea region (Alaska, Russia), and the Appalachians, as well as numerous domes in Precambrian terranes. They consist of a dome of gneiss intruded by younger granite and migmatite and mantled with sedimentary rock. These have been interpreted as a consequence of two distinct mountain-forming events, with the first producing the granite basement and the second deforming and melting this basement to produce the domes. Figure 3.2 shows an image of Goodenough Island in Papua New Guinea (Ollier and Pain 1980).

With regard to other metamorphic rocks, marbles are metamorphosed carbonate rocks and quartzites are metamorphosed sandstones. Quartzite landforms are discussed in Sect. 5.3.

3.4 Inselbergs and Bornhardts

Some of the most impressive granitic landforms are inselbergs. These are isolated hills which rise up sharply above the level of their surrounding plains. They tend to be

Fig. 3.2 Goodenough Island, Papua New Guinea, gneiss dome. 9° 20′ 45.53″ S 150° 15′ 50.24″ E. ©Google Earth. Scale bar = 10 km

composed of hard, resistant rocks, and granites are one such host, as is the case with the Namibian inselbergs which will be discussed later. However, they can also occur in other rock types as is the case with the Australian inselbergs of Uluru—arkosic sandstone—and Kata Tjutu—conglomerate (see Chap. 5). They occur in a very wide range of climatic zones and range in size from small hillocks to huge, towering eminences. In granite areas, Twidale (1982) identified three principal morphological types, namely domes, nubbins (or block- and boulder-strewn inselbergs) and castle koppies (castellated inselbergs). Boulder inselbergs are seemingly rather chaotic heaps of detached rocks, which generally mantle a more massive, bedrock-rooted core or rest on a low-angle platform. They are relatively small, measuring from 1 to 2 m to more than 10 m across. Castellated inselbergs tend to have more angular outlines and are typically composed of a massive lower part, which grades into an array of pillars, walls and clefts in the upper part. They commonly show a stepped appearance, with vertical faces separated by subhorizontal benches. The former relate to vertical fractures, the latter to horizontal ones. The typical height of castellated inselbergs is of the order of tens of meters. The most impressive inselbergs, however, are domes. Their characteristic features include steep, bare and upward-convex slopes and sharp piedmont angles. The heights of domed inselbergs are very variable. Some are as low as 20–30 m, while others may exceed 500 m.

Bornhardts are a special category of inselberg and are dome-shaped, steep-sided hills, usually built of massive igneous rocks such as granite, with bare convex slopes covered with only sparse amounts of talus and with flattened summit surfaces. They form due to differential weathering and erosion, in the course of which the surrounding less massive rock is eroded away leaving massive, sparsely-jointed compartments. Many may have formed through selective deep weathering followed by stripping of the saprolite. However, they can also emerge through gradual lowering of the surrounding terrain in the absence of deep weathering. They are structurally controlled landforms and the existence of massive rock compartments is a necessary factor for their development. They often have slope-parallel joints called sheeting joints. These may result from unloading. Gradual opening of joints promotes slope instability, so that rock slides and falls involving large masses of rock are common on bornhardts. Consequently, whereas their upper slopes are bare and talus-free, their surrounding foot slopes may be covered by big blocks and boulders derived from upslope. While *bornhardt* was originally used to sometimes denote a type of insleberg (literally island mountain—an isolated dome in an otherwise flat landscape), the term *bornhardt* is used in modern literature to refer to domed hills and mountains regardless of isolation; thus, not all bornhardts are inselbergs and not all inselbergs are bornhardts (Migoń 2004).

3.4.1 The Namib Inselbergs

A fundamental event in the history of the Central Namib was an early Cretaceous phase of major tectonic activity. This was a time when the continental breakup of West Gondwanaland, sea-floor spreading, hot-spot formation and plume activity occurred (Goudie and Eckardt 1999). There is a group of large and distinctive igneous complexes, often called the Damaraland Complexes (Milner et al. 1995), associated with volcanic and intrusive activity caused by the opening up of the South Atlantic, the separation of southern Africa from South America, and the presence of an important hot-spot track off the coast of Africa, the submarine Walvis Ridge.

Spitzkoppe, in west central Namibia, is one of the tallest, if not the tallest inselberg on Earth (Migoń 2010a) and is often referred to as the 'Matterhorn of Africa' (Fig. 3.3). The gently sloping rock-cut surfaces around the hills, which typically slope at less than 2°, are pediments, across which shallow ephemeral channels are cut. Grosse and Klein Spitzkoppe and the Pondok Mountains are dramatically prominent inselberg groups that are visible from as far as 100 km away. They rise abruptly and spectacularly from a planation surface developed across basement mica schists and gneisses. Grosse Spitzkoppe reaches an altitude of 1728 m, Klein Spitzkoppe of 1584 m and the Pondoks of 1628 m, while the surrounding plains have an altitude of 1000–1100 m. Weathering pits are extensively developed and unusually large (Goudie and Migoń 1997) as are tafoni, rock shelters (as at Bushman's Paradise), and natural arches. The southwestern face of Grosse Spitzkoppe also shows the effects of large-scale rock slides, with a chaotic spread of boulders, some as much as 30 m long, lying beneath an area from which a great slab, which was at least 50 m thick, has evidently been dislodged (Goudie and Viles 2015). Details of the Spitzkoppe granites are given by Haapala et al. (2007). They are of early Cretaceous age (as are those of Brandberg and Erongo), and were emplaced at depths of several km below the ground surface that existed at the time, indicating the substantial amount of erosion that has occurred since then. Spitzkoppe is composed of relatively resistant potassium-rich granite which is mechanically strong, with widely spaced fractures, and is poorly jointed.

In the last decade or so, cosmogenic nuclides have been used to calculate the rate at which the Spitzkoppe granites are eroding (Matmon et al. 2013). These have suggested that the inselbergs are being lowered very gradually—at about 1–2 mm per thousand years, but that cliff retreat is taking place more speedily—c 8 mm per thousand years.

Located some 80 km inland from Namibia's Skeleton Coast, the Brandberg is an isolated massif that rises to almost 2600 m above sea level (Fig. 3.4). Its summit, the highest in Namibia, is at 2579 m, and rises above the desert

(a)

(b)

Fig. 3.3 **a** Spitzkoppe, Namibia. 21° 49′ 36.22″ S 15° 11′ 27.36″ E. ©Google Earth. Scale bar = 2 km. **b** Spitzkoppe, Namibia (*Photo* A. S. Goudie)

(a)

(b)

Fig. 3.4 **a** Brandberg, Namibia. 21° 7′ 50.05″ S, 14° 33′ 33.33″ E. ©Google Earth. Scale bar = 10 km. **b** Brandberg, Namibia (*Photo* A. S. Goudie)

plain, which has a mean elevation of around 700 m above sea level. With an area of c. 420–450 km^2, its nearly circular stock of granite, over 20 km in diameter, is surrounded by a collar or skirt of Karoo sediments and lava remnants. The Karoo strata generally dip between 2° and 5° towards the complex but may steepen to as much as 50° close to the contact, having been dragged downwards by subsidence of the granite along the peripheral ring-fault.

Brandberg was a volcano 25 km in diameter and 2 km high that was active immediately after eruption of the Etendeka volcanic rocks at c 132 mya in the Cretaceous. Following cauldron subsidence, which took place on a ring fracture 25 km in diameter, a sequence of granite structures were emplaced into the volcano. The great dome, created by the intrusion of the plutonic material, has since been eroded of the great bulk of the Karoo strata of which it was comprised. Apatite fission track analysis by Raab et al. (2005) indicates that there may have been as much as 5 km of denudation.

Erongo is the largest of the Cretaceous plutons in the central Namib (Falster et al. 2018; Goudie and Eckardt 1999; Pirajno 1990). It has a mean diameter of c 35–40 km. Formed between 137 and 124 mya, it is of the same age as the Etendeka Lavas (see Sect. 4.7). It has an irregular outline, and is the only one of the plutons in which the roof strata are still largely intact. Rising around 1000 m above the surrounding plains, it reaches an altitude of 2305 m above sea level. As with the Brandberg, Karoo sediments crop out as a band along the edge of the mountains and dip between 5° and 20° toward their centre. Erongo is a central caldera structure consisting of lavas and pyroclastic materials. It was formed by cauldron subsidence, following which the granitic rocks were passively emplaced in the space provided by that subsidence. The main complex is surrounded by an olivine dolerite ring-dyke which is up to 200 m thick.

The granites that outcrop at Ameib produce a very striking landscape (Fig. 3.5), with large domes, arches, precariously balanced boulders (as at the Bulls Parties), and remarkable weathering pits, while deep recesses produced by weathering have created numerous shallow caves or rock shelters. Weathering pits, the largest examples of which may be between 10 and 20 m long and are large enough to show up on Google Earth, occur not only at Erongo but also in the vicinity of the Spitzkoppe. They are known to exist in many different climatic zones, and also occur on sandstones, (see Sect. 5.2.7) but those on the Namib granites are exceptional for their size, their large numbers, and also for their bare floors, though no satisfactory explanation of their emptiness has yet been offered (Goudie and Migoń 1997; Migoń and Goudie 2003). Locally pits can cover up to one third of the rock surface. Tafoni can also be very large—attaining dimensions of up to 12 m across, 3–4 m deep, and 3–4 m high.

Another Namibian inselberg, Vogelfederberg, is discussed in Goudie (2022, Sect. 4.2.5).

3.4.1.1 Zimbabwe

Zimbabwe has large areas of ancient igneous and metamorphic rocks. The Zimbabwe craton consists of 23 greenstone belts (zones of variably metamorphosed volcanic sequences with associated sedimentary rocks that occur within Archaean and Proterozoic cratons between granite and gneiss bodies) surrounded by felsic gneisses and granites, with ages from 3.6 to 2.6 Ga. The Murehwa granite, in northwest Zimbabwe, belongs to the Chilimanzi suite of monzogranites that were intruded into the Zimbabwe craton in the late Archaean, and which occupy over 50% of the exposed surface area of the Zimbabwe craton (Blenkinsop and Treloar 2001; Jelsma et al. 2001). The steep and massively jointed domes are some of the most impressive in Zimbabwe (Fig. 3.6).

The Matopos (Matobo) Batholith in southwestern Zimbabwe is bounded to the north by the Bulawayo Greenstone Belt, to the east by the Filabusi Belt, and to the south by the Gwanda and Antelope Greenstone Belts (Pye et al. 1984). Greenstone belts have been interpreted as having formed at ancient oceanic spreading centers and island arc terranes. They are primarily formed of volcanic rocks, dominated by basalt, with minor sedimentary rocks interleaving the volcanic formations. In most places a fringe of gneissic granite separates the Matopos Granite from the Greenstones. The gneissic granite has been interpreted as a contact facies formed by interaction of the Matopos Granite with the Greenstones.

A close relationship between rock type and general topography exists in the area. The Greenstones give rise to gently rolling vegetated hills, while the gneissic granites generally form low ground. Localized areas of more massive gneiss sometimes give rise to low whaleback outcrops of bare rock. The Matopos Granite is characterized by rugged relief with massive domes and koppies (Fig. 3.7) separated by valleys which have developed along major joints and faults. It is generally medium- to coarse-grained and sometimes porphyritic, but towards the margins gneissic foliation is particularly marked and there are inclusions of mafic material and gneissic granite. These marginal rocks give rise to fewer and less massive outcrops than those closer to the centre of the batholith.

3.4.2 Morocco

In the Anti-Atlas of southern Morocco there is another area of granite landforms at a location called Tafraoute (Fig. 3.8). The weathering features are typical of granite and include tors, koppies and many cavernous weathering

(a)

(b)

Fig. 3.5 **a** Erongo, Namibia. 21° 38′ 55.90″ S, 15° 42′ 22.16″ E. ©Google Earth. Scale bar = 20 km. **b** Erongo granite, Namibia (*Photo* A. S. Goudie). **c** Erongo. Weathering pits in granite. 21° 47′ 11.28″ S, 15° 39′ 44.17″ E. ©Google Earth. Scale bar = 100 m. **d** Erongo weathering pit in granite. Pole for scale. Note the absence of debris (*Photo* A. S. Goudie)

(c)

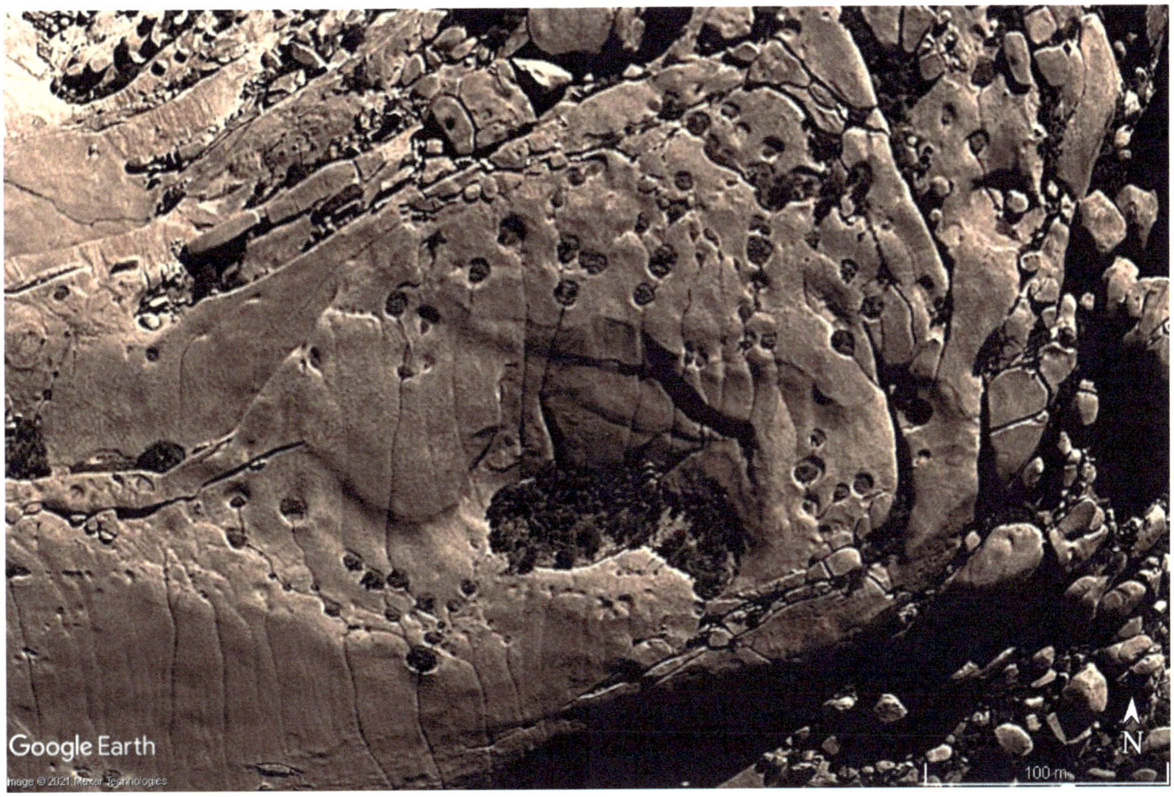

(d)

Fig. 3.5 (continued)

(a)

(b)

Fig. 3.6 **a** Murehwa (Mrewa), Zimbabwe, inselbergs. Note the importance of joints. 17° 37′ 13.63″ S, 31° 47′ 49.50″ E. ©Google Earth. Scale bar = 1 km. **b** Murehwa inselberg, Zimbabwe (*Photo* A. S. Goudie)

(a)

(b)

Fig. 3.7 **a** Matopos inselbergs, Zimbabwe. 20° 31′ 20.78″ S, 28° 25′ 23.57″ E. ©Google Earth. Scale bar = 200 m. **b** Matopos, Zimbabwe (*Photo* A. S. Goudie)

(a)

(b)

Fig. 3.8 **a** Tafraoute granite, Morocco. 29° 42′ 35.72″ N, 8° 59′ 20.63″ W. ©Google Earth. Scale bar = 70 m. **b** The Tafraoute landscape with tors and corestones (*Photo* A. S. Goudie). **c** Cavernous weathering forms in coarse granite at Tafraoute. This feature is reminiscent of a certain German Chancellor (*Photo* A. S. Goudie)

(c)

Fig. 3.8 (continued)

forms (tafoni). This is in a Cambrian mountain belt in which tectonic windows are found and in which Proterozoic rock complexes crop out. One of these windows is the Kerdous Massif in the western Anti-Atlas in which lies the Tasrirt Plateau at an altitude of 1000m (Soulaimani and Piqué 2004). The predominant rock in the area is the Tafraoute Granite. In the field, the 550 mya Tafraoute Granite is reddish-brown and rather coarse. The area is quite dry—precipitation is c. 250 mm per year—and salt weathering by sodium sulphate appears to be a major process operating on the granite (Schnepfleitner et al. 2016). Jointing is frequent rather than massive, so most features are relatively small.

3.4.3 Some American Examples

Although some local websites claim otherwise, Stone Mountain, Georgia, USA is not 'the biggest exposed granite on Earth'. Nonetheless it is an impressive dome of that material. Stone Mountain is located 25 km east of Atlanta (Fig. 3.9). It rises approximately 230 m above the surrounding countryside and covers 265 ha. It is elliptical in shape and trends in a northwest-southeast direction with its steepest side facing northward. The mountain occupies only a small portion of the area underlain by the granite, and, as Atkins and Joyce (1980, p. 1) remarked 'Stone Mountain may be compared to an iceberg in that only the tip of the granite body is above ground'. Approximately 300 mya Stone Mountain's granite was intruded at a depth of about 16 km below the surface. Weathering occurred, and slowly the layers of overlying rock above the granite were worn away, thus exposing the granite from about 15 mya. The granite is a primarily fine-grained quartz monzonite (Whitney et al. 1976) and the minerals within the rock include quartz, plagioclase feldspar, microcline, and muscovite, with smaller amounts of biotite and tourmaline. It is dotted with many weathering pits (Paradise and Yin 1993) and runnels.

Joshua Tree National Park in California has numerous relatively small hillocks of granite, similar to the koppies of Africa, together with rounded corestones (Fig. 3.10). They are developed in jointed Cretaceous monzonite (biotite-rich) granites (Trent 1998). As with so many granite areas

Fig. 3.9 Stone Mountain, Georgia, USA. 33° 48′ 24.65″ N, 84° 8′ 47.13″ W. ©Google Earth. Scale bar = 700 m

in the American south west, the slopes leading away from the hillocks are rock cut pediments. There is a sharp break in slope between the mountain front and the pediment, and this is called the piedmont angle (Fig. 3.11). Although pediments may develop on many rock types (Dohrenwend and Parsons 2009, pp. 390–391), the observed association of pediments with granitic rocks has often been argued as being due to the fact that granites break down discontinuously from boulders or corestones into constituent mineral grains. This was the view of Oberlander (1997, p. 151) who argued that 'on non-granite rocks the absence of a well-defined piedmont angle may reflect a continuum of clast sizes in transit from hillslope to pediment as opposed to the common bimodal distribution of granitic debris'.

3.5 Granite Weathering Forms

As we saw in the case of Erongo in Namibia and Tafraoute in Morocco, granite outcrops may display a range of weathering phenomena, including pits and tafoni, some of which may be of a considerable size.

An especially striking example of granites with runnels and karren comes from the dome of Pedra do Boca in north east Brazil (Fig. 3.12a). This has been described by Migoń and Maia (2020). The annual rainfall of the area is around 1000 mm. The igneous massif is a minor intrusion of latest Proterozoic age (c 573 mya) and is composed of granites, granodiorites and monzonites. Granites exposed within the locality are very coarse, porphyritic, with potassium feldspars as long as 5 cm. The largest domes in the region are more than 0.5 km long and 0.3–0.4 km wide, the smallest ones are c 150 × 150 m. The highest domes in the area have a relative height of c 200 m; others are in the range of 100–200 m. They have bare, convex slopes, some nearly vertical, and flattened summit surfaces. Several domes have impressive large caverns inset into their steep slopes. But it is the large number of runnels/karren which are ubiquitous on the flanks of the domes, which are especially interesting. Also in Brazil there are similar mega-karren on granites associated with the Quixadá Pluton (Souza et al. 2023; Santos et al. 2023) (e.g. at 5° 9′ 35.44″ S, 39° 5′ 10.05″ W).

In Madagascar, at Pic Boby in the Andringitra Massif, which is composed of syenite, there is a whole suite of weathering forms. These include caves, speleothems, basins (*vasques*), and, most strikingly of all, a series of runnels (*cannelures*) (Fig. 3.12b). Similar features can be found on inselbergs in northern Zimbabwe (e.g. at 17° 36′ 51.68″ S, 31° 48′ 12.26″ E).

(a)
(b)

Fig. 3.10 **a** Joshua Tree National Park, California. Granite inselberg and associated corestones. 33° 59′ 47.11″ N, 116° 3′ 34.80″ W. © Google Earth. Scale bar = 40 m. **b** A small granite inselberg in Joshua Tree National Park (*Photo* A. S. Goudie)

Fig. 3.11 Granite pediment in the Mojave Desert, California, USA. Note the sharp piedmont angle (*Photo* A. S. Goudie)

In the Seychelles, Neoproterozoic granites in the coastal region of Mahe, have fully developed systems of karren (Fig. 3.13). These are discussed by Johnson and Baarli (2005).

The deep weathering of granites can produce thick regolith (Fig. 3.14), and this is very important in terms of tor and inselberg formation (Linton 1955; Thomas 1966), but also for landslide generation in localities like Hong Kong (Durgin 1977; Ollier 2010), the Serra do Mar in Brazil (Picanço et al. 2014; Jaques et al. 2020) and Malaysia (Chigira et al. 2011). Figure 3.15 shows landslips produced in slopes developed on regolith on crystalline rocks near Nova Friburgo in Brazil in January 2011 (Netto et al. 2013).

Migoń (2010b) provided a good general assessment of the relationship between different grades of weathering and different types of mass movements in granite areas in the tropics, incuding Brazil and Hong Kong. He concluded (p. 43):

> Weathered terrains, especially in low latitudes, are characterized by frequent landslides, debris flows and earthflows, and cannot be adequately understood without considering the role of episodic gravity-driven processes. In basement terrains, where structural differentiation of weathering profiles is often fairly clear, there are clear relationships between weathering grade, relief and mass movement. Less weathered rock is able to support steeper terrain and hence debris slides turning not uncommonly into flows, sometimes followed by boulder movement, are frequent. In highly weathered terrain, dominated by clayey saprolites, the relief is more subdued and the dominant type of slope failure is rotational sliding, which may turn into earthflows

Some granite slopes are flared—they display concavities at their bases. A number of hypotheses have been advanced for their formation (Twidale and Harrell 2023). They have been interpreted as a two-stage landform shaped by etching during physico-chemical weathering in the shallow to deep subsurface, and subsequently exposed as a result of erosional stripping of either a weathered mantle or a sedimentary overburden. Others have seen them as the result of fluvial action or fire spalling (Buckman et al. 2021). The classic example is Wave Rock in Western Australia (Fig. 3.16).

(a)

(b)

Fig. 3.12 **a** Pedra da Boca inselberg with runnels, Brazil. 6° 27′ 17.04″ S, 35° 40′ 38.61″ W. ©Google Earth. Scale bar = 100 m. **b** Pic Boby, Andringitra Massif, Madagascar, showing runnel features developed in syenites. 22° 8′ 15.27″ S, 46° 51′ 39.63″ E. ©Google Earth. Scale bar = 100 m

Fig. 3.13 Mega-karren in granites in the Seychelles. The examples here are around 2 m high (*Photo* A. S. Goudie)

Many granite domes show exfoliation features (Reynolds et al. 2024). These have been related to the original cooling of the rock body, or to the decompression of the rock body as it is exhumed to the surface, or to processes at Earth's surface (such as thermal cycling and cracking). A good example is provided by Enchanted Rock, Texas (Fig. 3.17). This is the surface expression of a large Precambrian batholith called the Town Mountain Granite (Petersen 1984). In addition, the dome has many other weathering features (Pérez 2023).

3.6 Non-granite Inselbergs and Tors

Not all inselbergs, bornhardts, tors, and koppies are formed in granites. In Nigeria, the great Zuma Rock, which borders Abuja, is formed of an intrusive rock called gabbro. Other examples can occur in metamorphosed rocks, such as gneiss, migmatite and schist. Máčka et al. (2023), for example, describe gneissic tors from Czechia.

Perhaps the most famous of these gneissic features is the Sugar Loaf (*Pão de Açúcar*, in Portuguese) in Rio de Janeiro, Brazil (Fig. 3.18). Most of this 396 m high iconic landmark is composed, not of granite, but by Meso-Neoproterozoic high grade metasedimentary rocks—metapelitic gneiss (kinzigite) and augen gneiss. This gneiss is the product of metamorphism and deformation of a K-feldspar granite during the main collisional phase of the Ribeira orogenic belt at c 560 Ma (Ediacaran Period). There is also some later granite, which, for example, caps the Gávea Rock (Fernandes et al. 2010). Full details of the mineralogy of the gneiss are given in Castro et al. (2021).

The Kora National Park (formerly the home of George Adamson and his lions) lies almost on the Equator at the western edge of the eastern Kenya Plains on the Tana River, and consists largely of eroded Precambrian Basement rocks which in places are overlain by Late Cenozoic sediments (Pye et al. 1985). The most striking landforms are the domed inselbergs that have developed on some of the Basement System rocks (Fig. 3.19). Of the 100 inselbergs

Fig. 3.14 Deeply weathered granite regolith with core stones at Chapman's Peak on the Cape Peninsula, South Africa (*Photo* A. S. Goudie)

and other main outcrops of bedrock identified by air photo analysis, all but 12 occur in four areas of granitoid migmatite. Migmatites are metamorphosed igneous rocks, including granite.

The surrounding gneisses sometimes give rise to low ridges, but in general positive relief features are rare. Some outcrops of quartz-feldspar granoblastites and gabbro also form blocky, vegetated hills but do not give rise to inselbergs. The present landscape of the Kora area has evolved in response to tectonically-induced down-cutting by the Tana River during the late Tertiary and Quaternary. This incision is estimated to have occurred at an average rate of 0.07 mm/yr during the past five million years. The effect has been to lower the local base level for the sand rivers (*luggas*) flowing into the Tana, causing stripping of the regolith from the adjacent basement surface and thereby exposing fresh rock to sub-aerial weathering. The zone of regolith erosion has gradually extended north and south away from the Tana. During this process domed inselbergs appear to have developed on outcrops of granitoid migmatite which are more resistant to weathering and erosion than the surrounding gneisses. Samples of rock from the inselbergs were also found to be slightly richer in potassium than samples from the inter-inselberg areas (Pye et al. 1986). The available evidence suggests that this greater weathering resistance may be due to the relatively higher K-feldspar content of the migmatites.

Another example of landforms associated with metamorphic rocks are the angular tors of the Dunstan Range and other ridges in Otago on the South Island of New Zealand (Raeside 1949). These are composed of schists, which are part of the Otago Schist (Fig. 3.20). The rocks have been extensively metamorphosed and deformed and are part of the broader Haast Schist that occurs widely across New Zealand. Near Strath Taieri, the tors are often aligned, possibly representing bands of harder material in the schist. Their linear rows trend NW to NNW. The tors' typically columnar or blocky form is controlled by sub-vertical jointing and sub-horizontal schistosity. The cavernously weathered tor surfaces and weathering pits of up to 1 m in

Fig. 3.15 Landslides created by storm events, 2011, in Nova Friburgo, Brazil. 22° 15′ 9.31″ S, 42° 34′ 57.33″ W. ©Google Earth. Scale bar = 300 m

diameter (most are c 10–20 cm) on detached blocks around the base of many tors indicate that past periods of chemical weathering have been instrumental in their development (Fahey 1986). The in situ ^{10}Be ages give average surface ages of 122 ± 12 ka. This suggests that the majority of investigated tors have surfaced during the Marine Isotope Stage 5, which was one of the wettest and warmest climate periods (Raab et al. 2021), though during subsequent cold phases, periglacial processes such as solifluxion may have played a role in their exposure (Wood 1969; Fahey 1981). There may also have been some wind abrasion and erosion (Raab et al. 2022).

3.7 Conclusion

Landforms created by intrusive igneous rocks, such as granites, are among the most visited on Earth, but striking features may also be produced on metamorphic rocks such as migmatites, slates, phyllites, schists and gneisses. Theories as to the origins of certain groups of features, especially tors and inselbergs, have been the subject of considerable controversy, and have not been reviewed in depth here, but in-depth discussions occur in Migoń (2006) and Twidale and Romaní (2005).

(a)

(b)

Fig. 3.16 **a** Wave Rock, Hyden, Western Australia. 32° 26′ 42.97″ S, 118° 53′ 56.77″ E. ©Google Earth. Scale bar = 200 m. **b** The flared slope of Wave Rock, Western Australia (*Photo* A. S. Goudie)

Fig. 3.17 Exfoliation sheets on the sides of Enchanted Rock, Texas. 30° 30′ 28.52″ N, 98° 48′ 55.68″ W. ©Google Earth. Scale bar = 100 m

(a)

(b)

Fig. 3.18 **a** Sugar Loaf, Rio de Janeiro, Brazil. 22° 56′ 59.93″ S, 43° 9′ 35.86″ W. ©Google Earth. Scale bar = 700 m. **b** Sugar Loaf, Rio de Janeiro, Brazil (*Photo* A. S. Goudie)

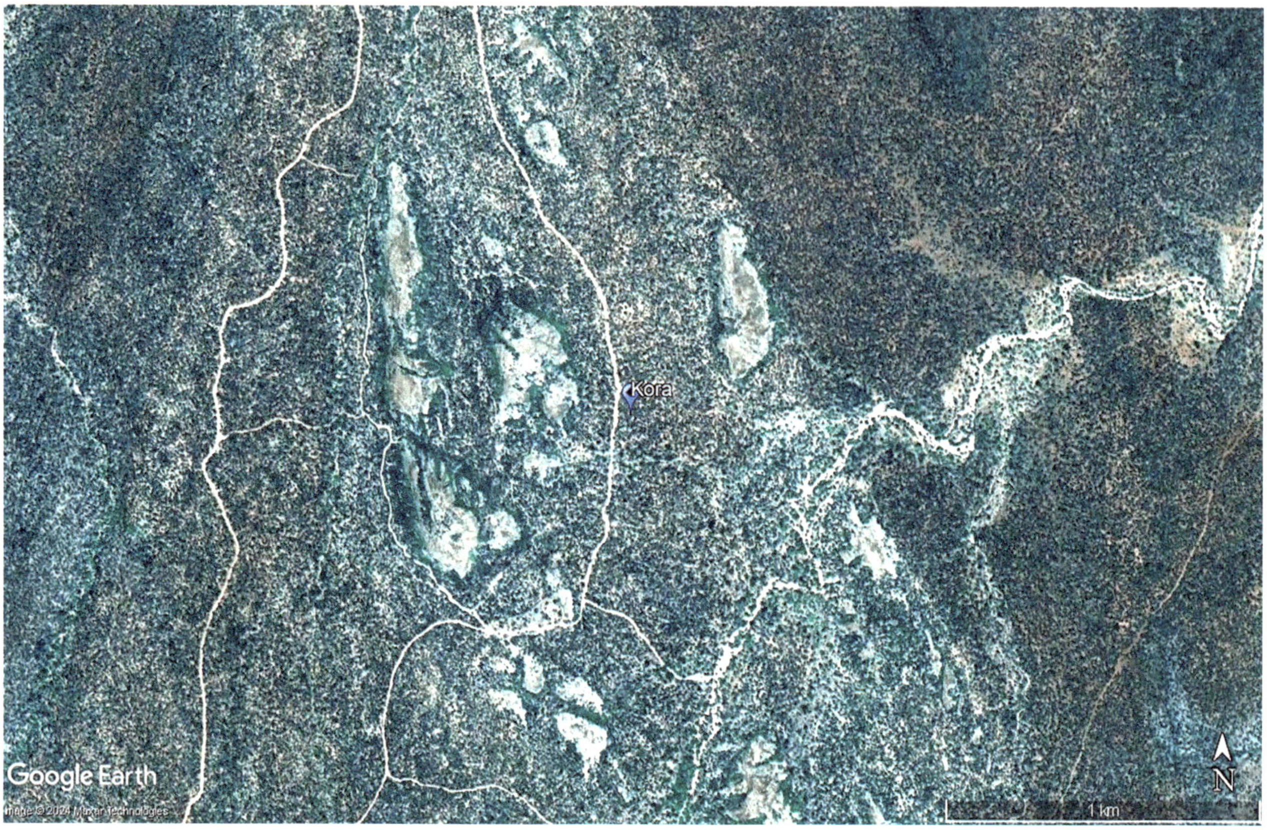

Fig. 3.19 **a** Kora inselbergs, Kenya. 0° 6′ 15.62″ S, 38° 45′ 50.78″ E. ©Google Earth. Scale bar = 1 km. **b** Kora inselberg, Kenya (*Photo* A. S. Goudie)

Fig. 3.19 (continued)

Fig. 3.20 Dunstan range schist tors, Otago, New Zealand. 45° 3′ 10.90″ S, 169° 22′ 23.16″ E. ©Google Earth. Scale bar = 100 m

References

Atkins R, Joyce L (1980) Geologic guide to Stone Mountain Park. Georgia Department of Natural Resources, p 46

Blenkinsop TG, Treloar PJ (2001) Tabular intrusion and folding of the late Archaean Murehwa granite, Zimbabwe, during regional shortening. J Geol Soc 158(4):653–664

Buckman S, Morris RH, Bourman RP (2021) Fire-induced rock spalling as a mechanism of weathering responsible for flared slope and inselberg development. Nature Comm 12(1):2150

Calcaterra D, Gili JA, Iovinelli R (1998) Shallow landslides in deeply weathered slates of the Sierra de Collcerola (Catalonian Coastal Range, Spain). Eng Geol 50(3–4):283–298

Castro NF, Mansur KL, de Oliveira Frascá MHB, Silva REC (2021) A heritage stone of Rio de Janeiro (Brazil): the facoidal gneiss. Episodes 44(1):59–74

Chigira M, Mohamad Z, Sian LC, Komoo I (2011) Landslides in weathered granitic rocks in Japan and Malaysia. Bul Geol Soc Malaysia 57:1–6

Cui K, Wu G, Wang X, Chen W (2017) Behaviour of slate following freeze–thaw and dry–wet weathering processes. QJ Eng Geol Hydrogeol 50(2):117–125

Dohrenwend JC, Parsons AJ (2009) Pediments in arid environments. In: Parsons AJ, Abrahams AD (eds) Geomorphology of desert environments, 2nd ed. Springer, Heidelberg, pp 377–341

dos Santos GLB, Maia RP, Travassos LEP, de Souza ASV (2023) Origin and evolution of solutional features on granitic inselbergs. Rev Brasil Geomorfol 25(1). https://doi.org/10.20502/rbgeomorfologia.v25i1.2469

Durgin PB (1977) Landslides and the weathering of granitic rocks. Geol Soc Am Rev Eng Geol 3:127–131

Evans DJA (2020) Bodmin Moor. In: Goudie AS, Migoń P (eds) Landscapes and landforms of England and Wales. Springer, Cham, pp 257–268

Fahey BD (1981) Origin and age of upland schist tors in Central Otago, New Zealand. NZ J Geol Geophys 24(3):399–413

Fahey BD (1986) Weathering pit development in the Central Otago mountains of southern New Zealand. Arc Alp Res 18(3):337–347

Falster AU, Simmons WB, Webber KL, Boudreaux AP (2018) Mineralogy and geochemistry of the Erongo sub-volcanic granite-miarolitic-pegmatite complex, Erongo, Namibia. Can Min 56(4):425–449

Fernandes NF, Tupinamba M, Mello CL, de OMN Peixoto (2010) Rio de Janeiro: a metropolis bdetween Granite-Gneiss massifs. In: Migoń P (ed) Geomorphological landscapes of the world. Springer, Dordrecht, pp 89–100

Gerrard J (1994) The landslide hazard in the Himalayas: geological control and human action. In: Proceedings of the 25th Binghamton symposium in geomorphology, SUNY, Binghamton, USA, pp 221–230

Goudie AS (2022) Desert landscapes of the world with Google Earth. Springer, Cham, p 271

Goudie AS, Eckardt F (1999) The evolution of the morphological framework of the central Namib Desert, Namibia, since the early Cretaceous. Geogr Ann 81A:443–458

Goudie AS, Migón P (1997) Weathering pits in the Spitzkoppe area, central Namib Desert. Z Geomorph 41:417–444

Goudie A, Viles H (2015) Weathering on the Namib plains: marble and granite. In: Goudie AS, Viles HA (eds) Landscapes and landforms of Namibia. Springer, Dordrecht, pp 91–96

Guan J, Wang Y, Li L, Niu L (2023) Geological challenges and stabilization strategies for phyllite rock slopes: a case study of Guang-Gansu expressway in Western China. Environ Sci Pollution Res 30(50):108741–108756

Gunnell Y, Jarman D (2020) Dartmoor. In: Goudie AS, Migon P (eds) Landscapes and landforms of England and Wales. Springer, Cham, pp 239–255

Haapala I, Frindt S, Kandara J (2007) Cretaceous Gross Spitzkoppe and Klein Spitzkoppe stocks in Namibia: topaz-bearing A-type granites related to continental rifting and mantle plume. Lithos 97:174–192

Jaques DS, Marques EAG, Marcellino LC, Leão MF, dos Santos F, Lemos CC (2020) Changes in the physical, mineralogical and geomechanical properties of a granitic rock from weathering zones in a tropical climate. Rock Mech Rock Eng 53:5345–5370

Jelsma HA, Becker JK, Siegesmund S (2001) Geochemical characteristics and tectonomagmatic evolution of late Archean granitoids from northern Zimbabwe. Z Deutsch Geol Gesell 152(2/4):199–226

Johnson ME, Baarli BG (2005) Erosion and burial of granite rocky shores in the recent and late Pleistocene of the Seychelles Islands: physical and biological perspectives. J Coast Res 21(5):867–879

Linton DL (1955) The problem of tors. Geogr J 121(4):470–487

Lo CM, Feng ZY (2014) Deformation characteristics of slate slopes associated with morphology and creep. Eng Geol 178:132–154

Máčka Z et al (2023) Gneissic tors in the central European upland: complex late Pleistocene forms? Geomorphology 436:108764

Matmon A, Mushkin A, Enzel Y, Grodek T, ASTER Team (2013) Erosion of a granite inselberg, Gross Spitzkoppe, Namib Desert. Geomorphology 201:52–59

Migoń P (2004) Bornhardt. In: Goudie AS (ed) Encyclopedia of geomorphology. Routledge, London, pp 92–93

Migoń P (2006) Granite landscapes of the world. Oxford University Press, Oxford, p 384

Migoń P (2010a) Spitzkoppe: the world of granite landforms. In: Migoń P (ed) Geomorphological landscapes of the world. Springer, Dordrecht, pp 155–162

Migoń P (2010b) Mass movement and landscape evolution in weathered granite and gneiss terrains. Geol Soc Lond Eng Geol Spec Pub 23(1):33–45

Migoń P, Goudie AS (2003) Granite landforms of the central Namib. Acta Univ Carolinae Geogr 35(Suppl):17–38

Migoń P, Maia RP (2020) Pedra da Boca, Pai Mateus, and Quixadá—three possible key geoheritage sites in Northeast Brazil. Geoheritage 12(3):51

Milner SC, Le Roex AP, O'Connor JM (1995) Age of Mesozoic igneous rocks in northwestern Namibia, and their relationship to continental breakup. J Geol Soc 152(1):97–104

Netto ALC et al (2013) January 2011: the extreme landslide disaster in Brazil. In: Margottini C, Canuti P, Sassa K (eds) Landslide science and practice: 6: risk assessment, management and mitigation. Springer, Berlin and Heidelberg, pp 377–384

Oberlander TM (1997) Slope and pediment systems. In: Thomas DSG (ed) Arid zone geomorphology, 2nd edn. Wiley-Blackwell, Chichester, pp 135–163

Ollier CD (2010) Very deep weathering and related landslides. Geol Soc London Eng Geol Special Publ 23(1):5–14

Ollier CD, Pain CF (1980) Actively rising surficial gneiss domes in Papua New Guinea. J Geol Soc Austral 27(1–2):33–44

Paradise TR, Yin ZY (1993) Weathering pit characteristics and topography on Stone Mountain, Georgia. Phys Geogr 14(1):68–80

Pérez FL (2023) Geoecology of a granite dome: Spatial interactions between gnammas, rills, soils, and plant cover, enchanted rock (Texas, USA). CATENA 223:106938

Petersen JF (1984) Enchanted rock state natural area: a guide to the landforms. Terra Cognita Press, San Marcos, p 56

Picanço J, Pinto CA, Mesquita MJ, Moraes M, Soares LF, Cardoso F (2014) Typology of rainfall-triggered landslides in the urban area of Antonina, Southern Brazil. In: Sassa K, Canuti P, Yin Y (eds) Landslide science for a safer geoenvironment, vol 3. Targeted landslides. Springer, Cham, pp 379–383

Pirajno F (1990) Geology, geochemistry and mineralisation of the Erongo Volcanic Complex. Namibia. S Afr J Geol 93(3):485–504

Pye K, Goudie AS, Thomas DSG (1984) A test of petrological control in the development of bornhardts and koppies on the Matopos Batholith, Zimbabwe. Earth Surf Proc Landf 9(5):455–467

Pye K, Goudie AS, Watson A (1985) An introduction to the physical geography of the Kora area, central Kenya. Geogr J 151:168–181

Pye K, Goudie AS, Watson A (1986) Petrological influence on differential weathering and inselberg development in the Kora area of central Kenya. Earth Surf Proc Landf 11(1):41–52

Raab MJ, Brown RW, Gallagher K, Weber K, Gleadow AJW (2005) Denudational and thermal history of the early Cretaceous Brandberg and Okenyenya igneous complexes on Namibia's Atlantic passive margin. Tectonics 24. https://doi.org/10.1029/2004TC001688

Raab G, Martin AP, Norton KP, Christl M, Scarciglia F, Egli M (2021) Complex patterns of schist tor exposure and surface uplift, Otago (New Zealand). Geomorphology 389:107849

Raab G et al (2022) Soil weathering dynamics and erosion in a dry oceanic area of the southern hemisphere (Otago, New Zealand). Sci Rep 12(1):19803

Raeside JD (1949) The origin of schist tors in central Otago. NZ Geogr 5(1):72–76

Reynolds A, Lang K, Arson C (2024) A comparison of exfoliation joint formation mechanisms: what is the role of surface processes? Copernicus Meetings, EGU24-1197. https://doi.org/10.5194/egusphere-egu24-1197

Schnepfleitner H, Sass O, Fruhmann S, Viles H, Goudie A (2016) A multi-method investigation of temperature, moisture and salt dynamics in tafoni (Tafraoute, Morocco). Earth Surf Proc Landf 41(4):473–485

Soulaimani A, Piqué A (2004) The Tasrirt structure (Kerdous inlier, Western Anti-Atlas, Morocco): a late Pan-African transtensive dome. J Afr Earth Sci 39(3–5):247–255

Souza AS, Maia RP, Bezerra FH, Migoń P, Siame LL (2023) Granitic inselberg erosion controlled by dike swarm array in semiarid Brazil. Geomorphology 440:108865

Sparks BW (1971) Rocks and relief. Longman, London, p 422

Teyssier C, Whitney DL (2002) Gneiss domes and orogeny. Geology 30(12):1139–1142

Thomas MF (1966) Some geomorphological implications of deep weathering patterns in crystalline rocks in Nigeria. Trans Inst Br Geogr 40:173–193

Trent DD (1998) Geology of Joshua tree national park. California Geol 51(5):1–28

Twidale CR (1982) The evolution of bornhardts: the nature of these dramatic landforms that rise abruptly from flat plains is now beginning to be more fully understood. Am Sci 70(3):268–276

Twidale CR, Harrell JA (2023) Flared slopes: the work of water or fire? Earth Sci Rev 247:104611

Twidale CR, Romaní JRV (2005) Landforms and geology of granite terrains. Balkema, Leiden, p 351

Whitney JA, Jones LM, Walker RL (1976) Age and origin of the Stone Mountain granite, Lithonia district, Georgia. Geol Soc Am Bull 87(7):1067–1077

Whitney DL, Teyssier C, Vanderhaeghe O (2004) Gneiss domes and crustal flow. Geol Soc Am Spec Paper 380:1–12

Wood BL (1969) Periglacial tor topography in southern New Zealand. NZ J Geol Geophys 12(2–3):361–375

Yokoyama O (2020) Evolution of uphill-facing scarps by flexural toppling of slate with high-angle faults. Geomorphology 352:106977

Volcanoes

4

Abstract

Volcanoes are sites on the surface of the planet through which gaseous, liquid and/or solid materials are expelled or erupted through the action of internal thermal processes. Volcanic eruptions produce two main types of material—lavas and pyroclasts. Among the major features associated with them are craters and calderas, maars, lava flows and plateaux, collapse features, and lahars (mudflows). Many volcanoes are associated with plate tectonics, being related to subduction zones, hot spots, etc.

Keywords

Volcano · Lava · Pyroclast · Caldera · Crater · Lahar · Maar · Passive margin · Great escarpment

4.1 Introduction

Volcanoes can be defined as sites on the surface of the planet through which gaseous, liquid and/or solid materials are expelled or erupted through the action of internal thermal processes. They are themselves landforms, but they have many different morphologies depending on their style of eruption and the materials of which they are composed (Bullard 1962; Ollier 1988; Scarth 1994; Thouret 1999; Francis and Oppenheimer 2004; Giordano et al. 2024). Some volcanoes are currently active, some are dormant and others are extinct. Volcanic eruptions produce two main types of material—lavas and pyroclasts (literally 'broken by fire'). The former tend to be predominantly effusive while the latter are explosive. Perhaps the simplest volcanic construct is the cinder or scoria cone. Shield volcanoes are usually crowned by nested and intersecting calderas and collapse pits formed by subsidence. When there is abundant ground water, hydrovolcanic eruptions can blast out wide depressions bounded by low, circular or elliptical rims of ejecta. These are called *maars*. Most volcanoes are polygenetic and are the result of numerous eruptive episodes. Simple cones (also called stratovolcanoes) are usually composed of interlayered lava and tephra (ash) produced in multiple eruptions over the life of the volcano. Ford (2023) has some outstanding ground and space images of volcanoes from all over the world.

4.2 Calderas and Craters

Calderas are large circular or elliptical volcanic depressions whose diameter (typically several or several tens of km) greatly exceeds those of any included vents. They are formed by the evacuation of a magma chamber within the crust, and subsidence of the overlying rocks. Calderas can form on volcanoes of different magma composition, from low silica (mafic) to high silica (silicic). Though there are no hard and fast distinctions, the term crater tends to be reserved for smaller features created as a result of excavation of rock during explosive eruptions or smaller-scale collapse (collapse pits). The highest magnitude explosive eruptions on Earth, sometimes called super-eruptions, have generated the largest calderas. Volcanoes that have experienced more than one super-eruption are colloquially known as super-volcanoes. Calderas can be complex, and obscured by post-collapse uplift, volcanism and erosion. They are enclosed by a topographic rim. This the head of the escarpment bounding the caldera. The inner walls can form cliffs in young calderas but they retreat through time as a result of landsliding. Rock redistributed by mass movements may form a collapse collar on the caldera floor.

A volcanic crater is a bowl- or funnel-shaped depression that usually lies directly above the vent from which

A. Goudie, *Rocky Landscapes of the World with Google Earth*, https://doi.org/10.1007/978-3-031-77418-8_4

volcanic material is ejected. Craters are commonly found at the summit of volcanic edifices, but they may form above satellite (flank) vents of composite and shield volcanoes.

Let us now consider a range of volcanic edifices from different parts of the world, starting with South America, for it has many volcanoes dotted along the Andes. These mountains are the result of the subduction of oceanic crust beneath the South American Plate as the Nazca and the South American plates converge. One of the Andean volcanoes is Cerro Galan (see Francis and Oppenheimer 2004, Sect. 11.2) (Fig. 4.1). This is a huge volcanic feature in northwest Argentina, which has a diameter of c 35 km. One unusual feature of the Cerro Galan system is its longevity. Biotite^{40}Ar/^{39}Ar ages indicated that the first major eruption of the Cerro Galan magmatic system occurred c. 5.6 mya. Its caldera seems to have been formed by a great eruption c 2.2 mya. Its summit lies at c 6600 m above sea level, while the floor of the caldera lies at c 4800 m. It is flanked by extensive ash deposits called ignimbrites (Sparks et al. 1985) of at least nine different ages (Folkes et al. 2011).

In the Sahara desert in North Africa, several rugged volcanic massifs are present. The most notable of these are Jebel Marra in Sudan, Tibesti in Chad (see Fig. 4.34), and Ahoggar in Algeria (Azzouni-Sekkal et al. 2007). In addition, in the Fezzan area of southern Libya, there is a black dot that makes a striking contrast to the generally light-coloured desert surface. This is the Waw an Namus (Fig. 4.2). It is a 100 m deep, 4 km wide, oval-shaped caldera (Bardintzeff et al. 2012). Its outer rim forms a low topographic feature in the landscape, but a steep margin inwards. Within the caldera lies a 140 m high and 1.3 km wide scoria cone. At the base of the cone, three small lakes containing salt-rich warm water are aligned along the caldera rim. Dark-coloured tephra of basaltic composition have buried the desert sediments around the caldera to distances of 10–20 km. The tephra deposit consists of volcanic ash and lapilli. It is this that produces the conspicuous colour contrast visible on Google Earth images. The orientation of the ash plume is the same as that of the neighbouring dunes, both of which have been the result of the north east trade winds.

The age of the volcano is still uncertain, but the central scoria cone may be only a few thousand years old, and may possibly even be of historical age. While the precise date of the most recent eruption is not known, the lack of erosion and weathering implies that it probably occurred in the last few thousand years.

Off North Africa lie the Azores. The Azores archipelago is composed of nine volcanic islands. These lie on a hot spot associated with the lateral branch of the Mid-Atlantic

Fig. 4.1 Cerro Galan, Argentina 25° 59′ 24.93″ S, 66° 52′ 54.83″ W. ©Google Earth. Scale bar = 10 km

Fig. 4.2 Waw an Namus, Libya, crater and ash plume. 22° 23′ 36.72″ N, 24° 22′ 24.30″ E. ©Google Earth. Scale bar = 3 km

Ridge near the junction of three major tectonic plates; the North American Plate, the Eurasian Plate and the African Plate. Sete Cidades (Fig. 4.3) is an active central stratovolcano on the western part of São Miguel island. Subaerial activity started more than 250 ka ago (Queiroz et al 2008, 2015). The present caldera developed in three phases associated with massive paroxysmal eruptions which occurred approximately 36,000, 29,000 and 16,000 years before present. Rising about 3 km from the average depth of the seafloor of the Azores Platform, the volcanic massif is dominated by an active central volcano, which has developed a roughly circular caldera at its summit. Sete Cidades was a composite edifice formed from the accumulation of pyroclastic deposits and lava flows. It originated from a central vent that produced numerous eruptions of variable style, magnitude and duration. This volcano has an average subaerial basal diameter of about 12 km and currently has a maximum altitude of 845 m. At its summit is a well-developed caldera whuich is roughly circular, with a diameter of about 5 km and an average depth of 350 m. It is bounded by steep sub-vertical walls, and at the foot of these walls thick accumulations of talus frequently occur. Several secondary volcanic cones (i.e. pumice and scoria cones, maars and domes) and four lakes occur inside the caldera. The outer flanks of the central volcano rise to 18° near to the caldera rim. On those upper slopes, where the thickness of loose pyroclastic materials is greatest, streams have incised deep valleys as they run down to the Atlantic Ocean, and form a near radial pattern around the summit caldera.

Similarly dramatic drainage patterns are found on the flanks of Mount Bromo, a small pyroclastic cone (Fig. 4.4). This is one of the most active volcanoes in Indonesia, located on the eastern part of Java Island. Volcanism in Java is related to subduction of the Australian Plate underneath the Eurasia Plate along the Sunda arc. Over the last two centuries, more than 50 eruptions have been recorded on this volcano. It is one of four volcanoes that occur within the larger and older Mount Tengger caldera (Van Gerven and Pichler 1995; Alfinati et al. 2022), the diameter of which is 9 km from north to south. Bromo's elevation is about 2392 m above sea level and 133 m above the base of Tengger Caldera. Its crater size is c 600 m × 800 m. Given the youthfulness of Bromo, it is evident that stream erosion and drainage incision occurs very in this environment.

Hot spots elsewhere in the oceans have also created volcanoes. The Hawaiian archipelago is part of a ridge of volcanic cones, stretching from Kure (northwest of Midway) to the island of Hawaii at the south east (Walker 1990). The islands were formed by a hot spot occurring in the middle of the Pacific Plate. While the hot spot itself is fixed, the

Fig. 4.3 Sete Cidades, Azores. 37° 51′ 40.33″ N, 25° 45′ 53.49″ W. ©Google Earth. Scale bar = 5 km. 1 = Seara, 2 = Caldeira do Alferes, 3 = Caldeira Seca, 4 = Lagoa Rasa, and 5 = Lagoa de Santiago. Note the radial drainage pattern

plate is moving. So, as the plate moved over the hot spot, the string of islands that make up the Hawaiian Island chain were formed. The age of the archipelago has been estimated using potassium-argon dating methods. From this study and others, it is estimated that the northwesternmost island, Kure Atoll, is the oldest at approximately 28 mya; while the southeasternmost island, Hawaii, is approximately 0.4 mya. Only on the island of Hawaii are there presently active volcanoes, though as Walker (1990, p. 319) remarked: 'Hawaiian volcanoes at the peak of their activity attain an output unsurpassed by any other volcano on earth and probably approached or matched only by volcanoes in Iceland, Reunion, and the Galapagos'.

Mauna Loa (4169 m) and Mauna Kea (4205 m) lie on the island of Hawaii (Fig. 4.5). Also important is the Kilauea Crater (1247 m). Radiating out from Mauna Loa are huge lava flows. On land, Mauna Loa has a crater approximately 15 km in circumference although the base of this volcano, 10,000 m below, is estimated at almost 500 km in diameter. The eruptions of low-viscosity lava, mainly from a central dome, spread out over large distances, creating a relatively gently sloping surface. This type of eruption is known as *Hawaiian* (Fig. 4.6). Thus Hawaiian volcanoes are of the broad shield type with slopes that typically range between 4 and 10° except where fault scarps produce much steeper slopes.

On Oahu, some of the volcanic lavas are deeply furrowed, and these furrows may be caused by high intensity rainfall, but solutional attack may have played a role in producing lapié-like forms (Palmer 1927) of enormous size (Fig. 4.7). The closely-spaced ravines give what has been described as 'washboard topography'.

Mayotte is a volcanic island in the Comoros archipelago between the eastern coast of Africa and the northern tip of Madagascar (Guilcher 1988; Berthod et al. 2021; Thinon et al. 2022). It consists of two principal islands: the main island (known as Grande Terre) and a smaller island to the east, known as Pamandzi or Petite Terre (Fig. 4.8a). Volcanic activity started about 7.7 mya in the south, ceasing about 2.7 mya. In the north, activity started about 4.7 mya and lasted until about 1.4 mya. Mayotte is a primarily volcanic island rising steeply from the bed of the ocean to a height of 660 m on Mont Bénara. The island resulted from the rifting of Madagascar away from Africa as well as 'hotspot' mantle plume activity activity, and is also impacted by seismicity and deformation associated with the East African Rift. It is now surrounded by a coral reef and lagoon. Multiple earthquakes shook Mayotte between May

Fig. 4.4 Mount Bromo and the Tengger volcano in Java, Indonesia, showing the incised, radial drainage pattern. 7° 57′ 7.02″ S, 112° 56′ 57.76″ E. ©Google Earth. Scale bar = 1 km

2018 and July 2019, and more than a dozen in that year were magnitude of 5 or more. Landslides followed these earthquakes.

Another volcano that has subsided and is surrounded by a coral reef is Bora Bora in French Polynesia (Gischler et al 2016). Darwin (1842) used Bora Bora as the type site to explain the genetic connection between shoreline-attached fringing reefs, barrier reefs (detached from the shore by a lagoon), and atolls (annular reefs with a deep central lagoon) (Fig. 4.8b).

In the West Indies there is an arc of islands called the Lesser Antilles. These occur on a subduction zone associated with the boundary of the Caribbean and South American plates. Examples include Mount Pelée on Martinique, the Soufrière Hills on Montserrat, and Nevis Peak in St. Kitts and Nevis (see Allen 2017). The island of Nevis is dominated by Nevis Peak. It rises to 985 m and is comprised of a central cone with flank deposits radiating out toward the sea, and two smaller lava domes (Fig. 4.9).

Mount Pelée (Fig. 4.10), on the West Indian island of Martinique, gives its name to a particular type of eruption known as *Peléan.* It is highly explosive, as shown by the great eruption of 1902–3, and involves the ejection of *nuée ardente.* This term was first introduced by Lacroix in 1903 to describe the mixtures of gases and pyroclastic material that flowed rapidly down the flanks of Mt. Pelée in its volcanic activity of 1902–3. A detailed discussion of the havoc wrought by the eruption, which destroyed the city of St Pierre, is given in Bullard (1962) and further information in given in Francis and Oppenheimer (2004, Sect. 3.3). The most violent form of eruption is termed *Plinian,* and includes that of Mount St. Helens in the USA in 1980 and Pinatubo in the Philippines in 1991. Plinian eruptions are highly explosive, propelling dense clouds of gas and tephra to considerable heights. The volume of the explosions is usually sufficiently great to disrupt the structure of a stratovolcano, leading to flank fissure eruptions, destruction of crater walls and fragmentation of solidified lava masses. Pyroclastic-fall deposits are those created by explosive volcanic eruptions that have accumulated as a result of gravitational fall.

On the Caribbean island of Montserrat lies the highly active Soufrière Hills stratovolcano (Fig. 4.11). The volcano is andesitic in nature. Since the 1990s it has sometimes been extremely violent (Nelson 2017), with dome formation and explosions of pyroclastic material. These flows of pyroclastic material are clearly evident as the grey lines on the Google Earth image. By August 1997, much of the southern

Fig. 4.5 Mauna Loa shield volcano, Hawaii, USA. 19° 32′ 45.24″ N, 155° 23′ 18.83″ W. ©Google Earth. Scale bar = 60 km

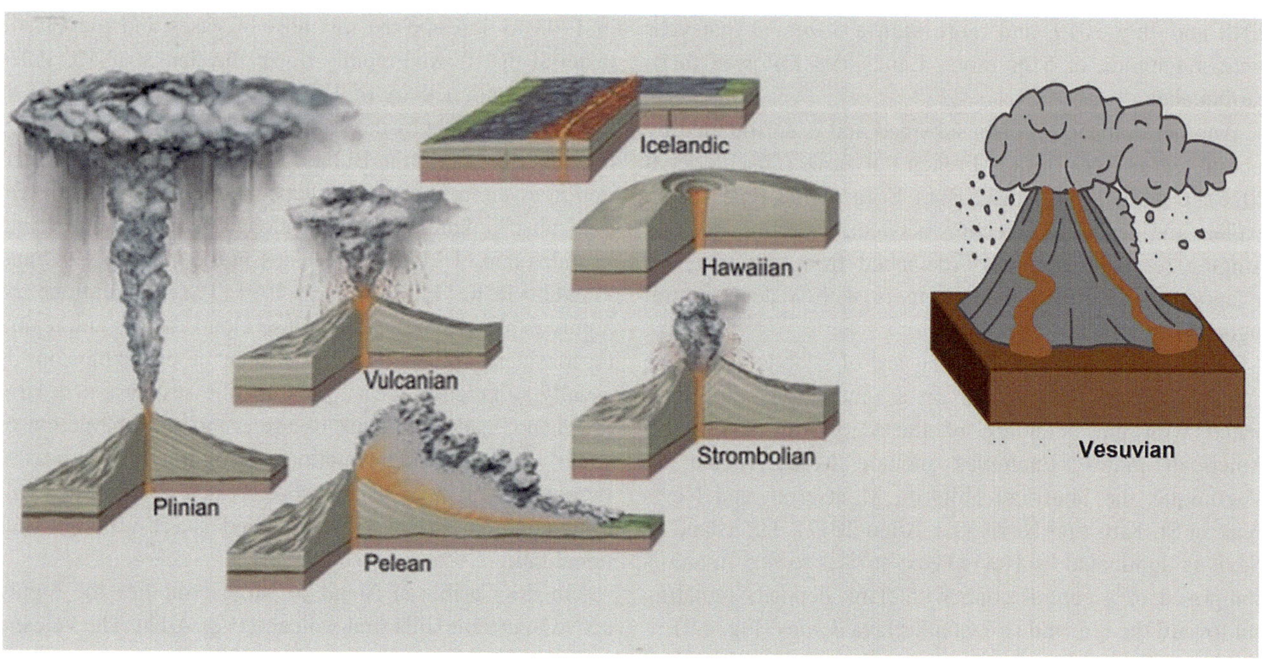

Fig. 4.6 Some major types of volcanic eruption. Modified after https://www.britannica.com/science/volcano/Six-types-of-eruptions (Accessed 17th August 2024) and https://www.alevelgeography.com/types-of-volcanoes/ (Accessed 17th August 2024)

Fig. 4.7 Oahu lava lapiés (?). 21° 31′ 11.75″ N, 157° 51′ 1.52″ W. ©Google Earth. Scale bar = 500 m

part of the island, including most of Plymouth, had been buried by pyroclastic flows (*nuées ardentes*).

Mount Pinatubo is in the Philippines. It is an active stratovolcano in the Zambales Mountains (Fig. 4.12). It was formed by the Eurasian Plate sliding under the Philippine Mobile Belt along the Manila Trench. Pinatubo is most notorious for its eruption in 1991, the second-largest terrestrial eruption of the twentieth century (Newhall and Punongbayan 1996). It ejected more particulates into the stratosphere than any eruption since that of Krakatoa in 1883 and caused world-wide climate change (Kinnison et al. 1994). So much magma was ejected in 1991 that the summit collapsed to form a caldera. High monsoonal rainfall soon filled this up to create Lake Pinatubo. There were also numerous lahars (Van Westen and Daag 2005).

On Sumatra there is a caldera that is one of the most striking on Earth, measuring 100 by 30 km. This steep-walled caldera contains Lake Toba, the largest volcanic lake on the planet (Fig. 4.13). It is a UNESCO Global Geopark. It is surrounded by steep cliffs, 400 to 1,200 m above the lake surface, and the water has a maximum depth of 585 m. Erupting at least 2800 km^3 of magma in a single event 74,000 years ago, it was created by what ranks as the largest volcanic eruption of the Quaternary. As a consequence, a vast portion of North Sumatra was smothered in ignimbrite that extended from coast to coast, an extensive ash deposit blanketed large portions of Southeast Asia, including the entire subcontinent of India, and an aerosol cloud and dust cloud probably encompassed the Earth (Chesner 2012). This volcanic activity was associated with subduction down the Sunda Trench of the Indo-Australian plate beneath the continental Eurasian (Sundaland) plate.

New Zealand is another location with much volcanic activity. This is primarily due to the country's position on the boundary between the Indo-Australian and Pacific plates, a part of the so-called 'Pacific Ring of Fire', and particularly the subduction of the latter under the former. One superb example of volcanic activity on the North Island is a steep stratovolcano called Mount Tongariro (Fig. 4.14). Other dramatic locations are Mount Taranaki (Egmont) and Mount Ruapehu. Andesitic eruptions formed Tongariro, which reaches a height of 1978 m. It is composed of layers of both lava and tephra and the eruptions that built the current stratovolcano commenced about 275,000 years ago (Leonard et al. 2021). It has been active in recent years and eruptions are now thought to have occurred in 1869, 1892, 1896–97, 1899, 1926, 1927, 1934 and 2012 (Scott and Potter 2014).

On the South Island, the Banks Peninsula comprises the eroded remnants of two large stratovolcanoes, Lyttelton,

Fig. 4.8 **a** Mayotte, Indian Ocean, with surrounding coral reef. 12° 49′ 43.30″ S, 45° 7′ 20.41″ E. ©Google Earth. Scale bar = 20 km. **b** Subsided volcano and reef, Bora Bora, French Polynesia. 16° 29′ 59.84″ S, 151° 44′ 0.44″ W. ©Google Earth. Scale bar = 8 km

Fig. 4.9 Nevis Island. 17° 9′ 3.51″ N, 62° 34′ 49.18″ W. ©Google Earth. Scale bar = 5 km

Fig. 4.10 Mt Pelée, Martinique. 14° 48′ 10.38″ N, 61° 9′ 21.39″ W. ©Google Earth. Scale bar = 4 km

Fig. 4.11 Soufrière Hills with pyroclastic flows, Montserrat. 16° 42′ 44.54″ N, 62° 10′ 30.98″ W. Scale bar is 3 km

Fig. 4.12 Mount Pinatubo crater lake, Philippines. 15° 8′ 35.75″ N, 120° 21′ 4.05″ E. ©Google Earth. Scale bar = 1 km

Fig. 4.13 Lake Toba, Sumatra. 2° 36′ 27.24″ N, 98° 57′ 55.80″ E. ©Google Earth. Scale bar = 40 km

Fig. 4.14 Mount Tongariro, New Zealand. 39° 8′ 23.64″ S, 175° 39′ 47.31″ E. ©Google Earth. Scale bar = 4 km

Fig. 4.15 Banks Peninsula, New Zealand. 43° 46′ 54.06″ S, 172° 52′ 17.45″ E. ©Google Earth. Scale bar = 10 km

which formed first, and Akaroa (Fig. 4.15). These formed by intraplate volcanism c 8 to 11 mya ago in the Miocene. The peninsula formed as offshore islands, with the volcanoes reaching to about 1,500 m above sea level. The two dominant craters were eroded, then flooded, to form the Lyttleton and Akaroa Harbours (Hampton and Cole 2009).

Hunga Tonga is a submarine volcano in the South Pacific and is located about 65 km north of Tonga's main island. It is part of the highly active Kermadec-Tonga subduction zone and its associated volcanic arc, which extends from New Zealand north-northeast to Fiji, and, as in New Zealand, is formed by the subduction of the Pacific Plate under the Indo-Pacific Plate. It has been highly active in the last two decades, but has also been subject to rapid marine erosion. Compare the morphology of the island (s) in 2009 and in 2019. They were changed further by a major eruption in January 2022 (Manneela and Kumar 2022; Proud et al. 2022) (Fig. 4.16). Similarly great coastal erosion of volcanic islands has been witnessed at the Zubair group in the Red Sea (Xu et al. 2015) (15° 5′ 3.19″ N, 42° 9′ 7.27″ E).

In East Africa there are great rift valleys associated with the breakup of the African Plate. In this region there is a great deal of volcanic activity which produces such great mountains as Kilimanjaro and Mount Kenya. One beautiful example is Oi Doinyo Lengai in northern Tanzania (Fig. 4.17). This has been active since the Late Pleistocene and significant eruptions have occurred throughout the Holocene. This active volcano consists of a volcanic cone with two craters, the northern of which has erupted during historical time (including 2004, 2007, 2008 and 2013). It is a symmetrical cone that rises more than 1800 m above the surrounding rift valley. It is currently the only place on Earth that can erupt natrocarbonatite lava, which unlike ordinary lavas is crammed full of calcium, sodium, carbon dioxide and potassium rather than silicates. This makes the lava exceptionally runny (Ford 2023, p. 67).

Another huge crater in East Africa is that at Ngorongoro in the Gregory Rift (Fig. 4.18)—a UNESCO Global Geopark. This is said to be the world's largest inactive, intact and unfilled volcanic caldera (Żaba and Gaidzik 2011). The crater, formed when a great volcano erupted and collapsed on itself two to three mya, is 610 m deep and its floor covers 260 km^2. It has an internal diameter of c 22 by 18 km, and the internal walls have a maximum height of c 350 m. Estimates of the height of the original volcano range from 4500 to 5800 m. Within the crater is the alkaline Lake Magadi. A detailed map of the crater appears in Scoon (2021, p. 126).

Lake Shala is the deepest lake (257 m) and the largest crater (c 12 × 15 km) in the Main Ethiopian Rift (Fig. 4.19).

Fig. 4.16 Hunga Island, Tonga in 2009 (top), 2019 (middle) and 2022 (bottom). 20° 32′ 52.82″ S, 175° 23′ 31.12″ W. ©Google Earth. Scale bar = 1 km

Fig. 4.17 Oi Doinyo Lengai, Tanzania. 2° 45′ 44.28″ S, 35° 55′ 1.23″ E. ©Google Earth. Scale bar = 1 km

The Shala basin, whuch contains Lake Shala, is a caldera which probably collapsed during the late Pliocene (about 3–4 mya) following large eruptions of ignimbrites and pumice (Le Turdu et al. 1999). In the Late Pleistocene a mega-lake connected three Ethiopian rift valley lakes (Grove et al. 1975). That large body of water, called Lake Galla, began to contract about 2000 years ago. Lake Galla had declined so much that it became three separate bodies of water: Lakes Shala, Abijata, and Langano.

The Canary islands, off the northwest coast of Africa, are notable for some striking volcanoes, many of which are active. The largest of these is Mount Teide on the island of Tenerife (Fig. 4.20). Its summit is at 3718 m. If measured from the ocean floor, its height of 7500 m makes it the third-highest volcano in the world. Teide started forming around 170,000 years ago and its last eruption was in 1909. It is a stratovolcano formed by successive eruptions of lavas and pyroclastics (Carracedo et al. 2007). It has been the site of some major flank collapses since it was formed (Boulesteix et al 2012), and on the island there are three large amphitheatres—the valleys of Güimar, La Orotava and Icod. The most accepted theory suggests that large landslide events were the origin of these landforms (Hürlimann et al. 2004). Full details of the volcanic edifices of Tenerife are given in Criado and Paris (2005).

Lanzarote is another volcanic island in the Canaries, with a great deal of relatively recent activity producing fresh-looking calderas (as at Montana Blanca) (Fig. 4.21) and great expanses of lava. Particularly great eruptions occurred in the 1730s and buried about one quarter of the island (see Scarth 1994, pp. 121–132). Full details of the nature and history of the Lanzarote volcanics are given in Machin and Torrado (2005).

There are a number of volcanoes in Italy and neighbouring parts of the Mediterranean Sea. Etna is the largest volcano in Europe, and is c 50 km in diameter (Fig. 4.22). It dominates eastern Sicily, being over 3300 m high and covering an area of c 1200 km^2. It is situated on an uplifted fault block, part of a complex series of intersecting faults that run through Sicily and Calabria. Submarine eruptions, representing the earliest volcanic events in the Etna region, occurred about 500 ka ago, but the oldest subaerial volcanic products in the Etna region erupted about 330 ka ago (Branca et al. 2017). Finally, the eruptive activity of the past 60 ka has produced its present day morphology. Large lava flows radiate from its central cone and this volcano is still highly active with steam and gas emissions occurring at the central vent (Chester et al. 2012).

One of the volcanoes in the Lipari Islands, to the west of mainland Italy, is Stromboli (Fig. 4.23). It is the

(a)

(b)

Fig. 4.18 **a** Ngorongoro Crater and Lake Magadi, Tanzania. 3° 9′ 53.31″ S, 35° 35′ 43.63″ E. ©Google Earth. Scale bar = 9 km. Ngorogoro Crater and Lake Magadi (*Photo* Alice Goudie)

Fig. 4.19 **a** Shala Lake, Ethiopia. 7° 28′ 20.48″ N, 38° 30′ 31.26″ E. ©Google Earth. Scale bar = 10 km. **b** Shala Lake, Ethiopia (*Photo* A.S. Goudie)

Fig. 4.20 Teide, Tenerife, Spain, showing thr sites of three major flank failures. 28° 16′ 35.22″ N, 16° 37′ 49.43″ W. ©Google Earth. Scale bar = 3 km

northernmost island of the Aeolian Archipelago and represents only the uppermost portion (924 m above sea level) of a much larger edifice extending to a depth of 1500–2600 m below sea level (Bertagnini et al. 2011). It is a stratovolcano that has grown above the floor of the Mediterranean by the accumulation of lava and pyroclastics. It is is famous for its persistent activity that began in the present form between the third and seventh centuries AD. Persistent 'normal' activity consists of rhythmic, short-lived, mildly energetic explosions that eject ash- to bomb-sized fragments to heights of a few hundred metres above the craters. Sporadic lava flows occur either from the summit craters or from eruptive fractures within the Sciara del Fuoco, a horseshoe-shaped structure on the NW side of the volcano that developed through a series of slope failures (Lucchi et al. 2017). The persistent normal activity of Stromboli is occasionally interrupted by sudden and highly energetic explosive events called 'strombolian paroxysms'. It has given its name to a type of eruption known as *Strombolian*, which is remarkable for the regularity and frequency of basaltic eruptions of mild to moderate explosiveness.

Etna is another example of a Strombolian volcano, as is Paricutin in Mexico (see also Fig. 4.30). Where explosions are more violent than a volcano of the Strombolian type, they are termed *Vulcanian* (after Vulcano), which is also located in the Lipari Islands. Even more violent is the *Vesuvian* type of eruption, named after Mount Vesuvius near Naples. A catastrophic eruption in AD 79 buried the cities of Pompeii and Herculaneum.

A further example of a Strombolian volcano is Mount Erebus (Sims et al. 2021) (Fig. 4.24). This is the highest active volcano in Antarctica, and the southernmost active volcano on Earth. It has a summit elevation of 3794 m. It is the current eruptive centre of the Erebus hotspot. The summit contains a persistent lava lake. Characteristic eruptive activity consists of Strombolian eruptions from the lava lake or from one of several subsidiary vents, all within the volcano's inner crater.

The prosaicly-named Crater Lake is in south-central Oregon, USA. The lake partly fills a 655 m deep caldera that was formed by the collapse of the volcano Mount Mazama, which originally may have been 3600 m high

Fig. 4.21 Montana Blanca Caldera, Lanzarote, Spain. 29° 2′ 24.17″ N, 13° 43′ 48.17″ W. ©Google Earth. Scale bar = 1 km

(Fig. 4.25). The lake is 8.0 by 9.7 km across, with a caldera rim ranging in elevation from 2100 to 2400 m, and an average lake depth of 350 m. Mount Mazama, part of the Cascade Range volcanic arc, was built up mostly of andesite, dacite and rhyodacite over a period of at least 400,000 years. The caldera was created in a massive eruption between 6000 and 8000 years ago that led to the cauldron subsidence of Mount Mazama (see Francis and Oppenheimer 2004, Sect. 11.1; Bacon et al. 2002). Later volcanic activity led to the formation within the lake of Wizard Island, a small cinder cone.

Santorini caldera is a large, mostly submerged feature located in the southern Aegean Sea, 120 km north of Crete (Druitt et al. 1999) (Fig. 4.26). Visible above water is the circular Santorini island group, consisting of the main island of Santorini (in classic Greek called *Thera*), Therasia (Thirasia), and Aspronisi at the periphery, and the Kameni islands at the centre. The caldera measures about 12 by 7 km, with 300 m high steep cliffs on three sides. This complex is the most active part of the South Aegean Volcanic Arc. It has been formed by the subduction of the African Plate underneath the Aegean sub-plate of the Eurasian plate. The volcanism of Santorini, which commenced with submarine activity c 0.65 mya, is characterised by periodic Plinian eruptions. The current caldera was formed about 3600 years BP during the Minoan (Bronze Age) eruption (Friedrich et al. 2006).

4.3 Necks, Diatremes, Lava Lakes and Lava Tubes

Volcanic necks and diatremes (gas pipes) are both pipe-like masses of lava, agglomerates and breccias that occupy the former conduits for ejectamenta to reach Earth's surface. In many cases, because of their resistance, they stand up as a neck, plug, crag or tower. Arthur's Seat in Edinburgh, Scotland, is a famous example, as are the Hopi Buttes in Arizona. Ship Rock in New Mexico is a volcanic neck with prominent radiating dikes (see Sect. 2.5).

Devils Tower, in Wyoming, USA, is a 264 m high feature that arises abruptly from the prairies surrounding the Black Hills (Fig. 4.27). It has been proposed that it is a stump of magma that had been intruded into surrounding sedimentary rock, which was subsequently eroded away (Robinson 1956). The columnar structure of the magma is very plain to see. However, this has been disputed as an explanation by Závada et al. (2015) who believe Devils Tower is a remnant of a low lava dome that was emplaced into a broad phreatomagmatic (i.e. a type of explosive

(a)

(b)

Fig. 4.22 **a** Mount Etna, Sicily, Italy. 37° 45′ 14.85″ N, 14° 59′ 54.98″ E. ©Google Earth. Scale bar = 10 km. **b** Mount Etna, Sicily, Italy (*Photo* A.S. Goudie)

Fig. 4.23 Stromboli, Italy. 38° 47′ 34.68″ N, 15° 13′ 7.95″ E. ©Google Earth. Scale bar = 3 km

Fig. 4.24 Mt Erebus, Antarctica. 77° 31′ 38.02″ S, 167° 11′ 7.16″ E. ©Google Earth. Scale bar = 10 km

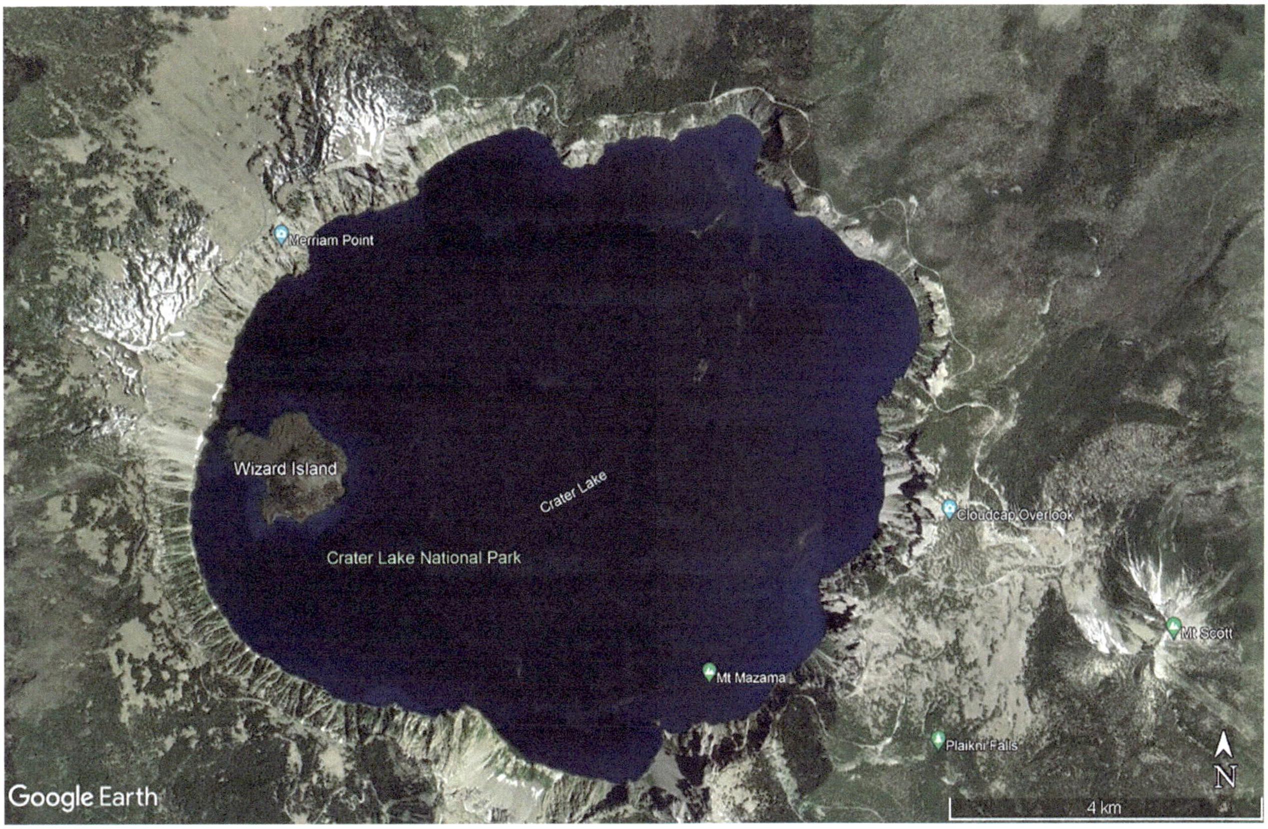

Fig. 4.25 Crater Lake Oregon, USA. 42° 56′ 3.83″ N, 122° 6′ 4.14″ W. ©Google Earth. Scale bar = 6 km

Fig. 4.26 Santorini, Greece. 36° 24′ 38.29″ N, 25° 26′ 3.88″ E. ©Google Earth. Scale bar = 9 km

Fig. 4.27 Devils Tower, Wyoming, USA. 44° 35′ 25.31″ N, 104° 42′ 52.96″ W. ©Google Earth. Scale bar = 100 m

eruption that results from magma erupting through water) crater at the top of a maar-diatreme volcano.

When low viscosity lavas erupt within a crater they tend to be confined within the vent region and crater, forming a lava pond or lake. A spectacular and long-lived example occurs in the Danakil Depression of Ethiopia (Oppenheimer and Yirgu 2002; Moore et al. 2019) in the shape of the Erta 'Ale volcano, which lies on the Nubia-Arabia extensional boundary, an incipient mid-ocean ridge (Fig. 4.28).

Lava tubes are conduits formed by flowing lava from a volcanic vent that moves beneath the hardened surface of a lava flow (Sauro et al. 2020). They have been found on Earth, Mars and the Moon. When the supply of lava ceases at the end of an eruption or the lava is diverted elsewhere, lava in the tube flows out and leaves a partially empty tunnel or cave beneath the ground. In due course, collapses may take place into these tubes, creating depressions that are visible on the ground surface. Lava tubes can be up to 14–15 m wide, and run anywhere from 1–15 m below the surface. They can also be extremely long. One tube from the Mauna Loa volcano on Hawaii, dating to 1859, enters the ocean about 50 km from its eruption point. Excellent examples of collapse depressions associated with lava tubes are known from the Undara area of Queensland in Australia (Fig. 4.29). The Undara volcano erupted 190,000 years ago and is part of the McBride Basalt Province which forms part of the East Australian Volcanic Zone (Atkinson et al. 1975; Whitehead 2010; Middleton et al. 2023). The Undara lava tube system, one of the longest in the world, can be clearly located on aerial photographs and satellite images. It stands out because most of the collapse depressions support rainforest type vegetation which contrasts sharply with the open forest of the surrounding country.

4.4 Cinder Cones

A cinder cone (or scoria cone) is a steep conical hill composed of pyroclastic fragments, such as clinkers, ash, or scoria that accumulated around a vent. They are perhaps the world's most common volcanic landform (Wood 1980). Scoria are irregularly-shaped, highly vesicular (bubble-shaped cavities) fragments of lava that are erupted into the air and are typically solid when they land. The pyroclastic fragments are formed by explosive eruptions from a single, typically cylindrical, vent. As the gas-charged lava is blown violently into the air, it breaks into small fragments that solidify and fall around the vent to form a cone that often is symmetrical with a nearly circular ground plan. Slopes tend to be between 30 and 40° and the cones range in size from

Fig. 4.28 Erta 'Ale lava lake, Danakil, Ethiopia. 13° 36′ 23.21″ N, 40° 39′ 43.93″ E. ©Google Earth. Scale bar = 200 m

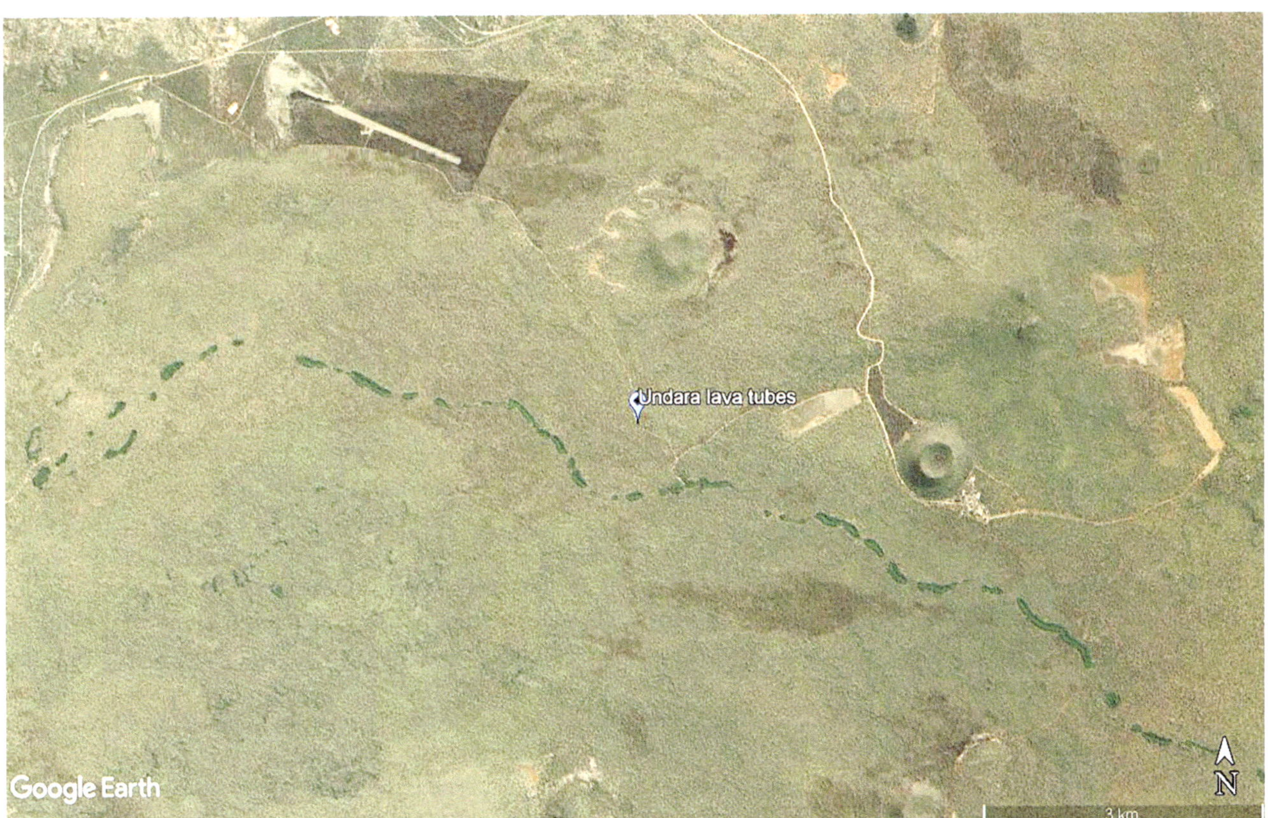

Fig. 4.29 An extremely long line of lava tube collapse depressions at Undara in Queensland, Australia. 18° 13′ 7.85″ S, 144° 38′ 13.94″ E. ©Google Earth. Scale bar = 3 km

Fig. 4.30 Parícutin, Mexico – a cinder cone. 19° 29′ 36.02″ N, 102° 15′ 2.88″ W. ©Google Earth. Scale bar = 400 m

tens to hundreds of meters tall. Most cinder cones have a bowl-shaped crater at the summit. Cinder cones are also commonly found on the flanks of shield volcanoes, calderas and stratovolcanoes. They include both monogenetic (single eruption) and polygenetic (multiple eruptions) cones, and may occur as isolated single cones or in fields. Because they are constructed of poorly consolidated, in some cases lightly sintered material, they are very erodible. One very famous cinder cone, called Parícutin, grew out of a Mexican corn field in 1943. It is located in Trans-Mexican Volcanic Belt in the Mexican state of Michoacán. Eruptions, which continued for nine years, built the cone to a height of 424 m (Pioli et al. 2008) (Fig. 4.30).

4.5 Maars and Tuff Rings

Maars are a type of what are called phreatic eruptions. These occur when a volcanic vent interacts with groundwater in the phreatic zone and releases vapour and fragments of surrounding rock (Schmincke 2007). The Eifel district of Germany is home to 72 of these. They are typically closed depressions between 100 and 3000 m wide and 10–500 m deep (Fig. 4.31a). They stand out from the general land surface and are composed primarily of a ring of fragments and lavas with slopes that dip 10–15° toward the outside of the crater and more steeply to the interior. Other examples of maars occur in many other locations including from Espenberg in Alaska (Fig. 4.31b). Located on the northern Seward Peninsula of Alaska these were formed by a series of Pleistocene basaltic eruptions through thick permafrost. The maars were excavated as much as 300 m into older lithologies; ranging from 4 to 8 km in diameter, they are the four largest known maars on Earth (Begét et al. 1996).

Whereas maars are excavated into the substrate, tuff rings and cones are built up above it. They are formed when magma comes near to the surface before being explosively fragmented. Good examples include Cerro Xico near Mexico City and Diamond Head on Oahu (Fig. 4.32). The latter, which is c 1 km in diameter, is asymmetric. Details are given in Wohletz and Sheridan (1983) and early discussions as to its controversial origin are provided by Hitchcock (1906), who believed the tuffs were thrown up explosively from beneath the level of the sea. For the most part the Head's rim is c 100 m high, but on its south western extremity it rises to 232 m (see elevation cross-profile). This was because the ejecta were moved by north easterly trade winds which inevitably led to thicker accumulation on the downwind side.

(a)

(b)

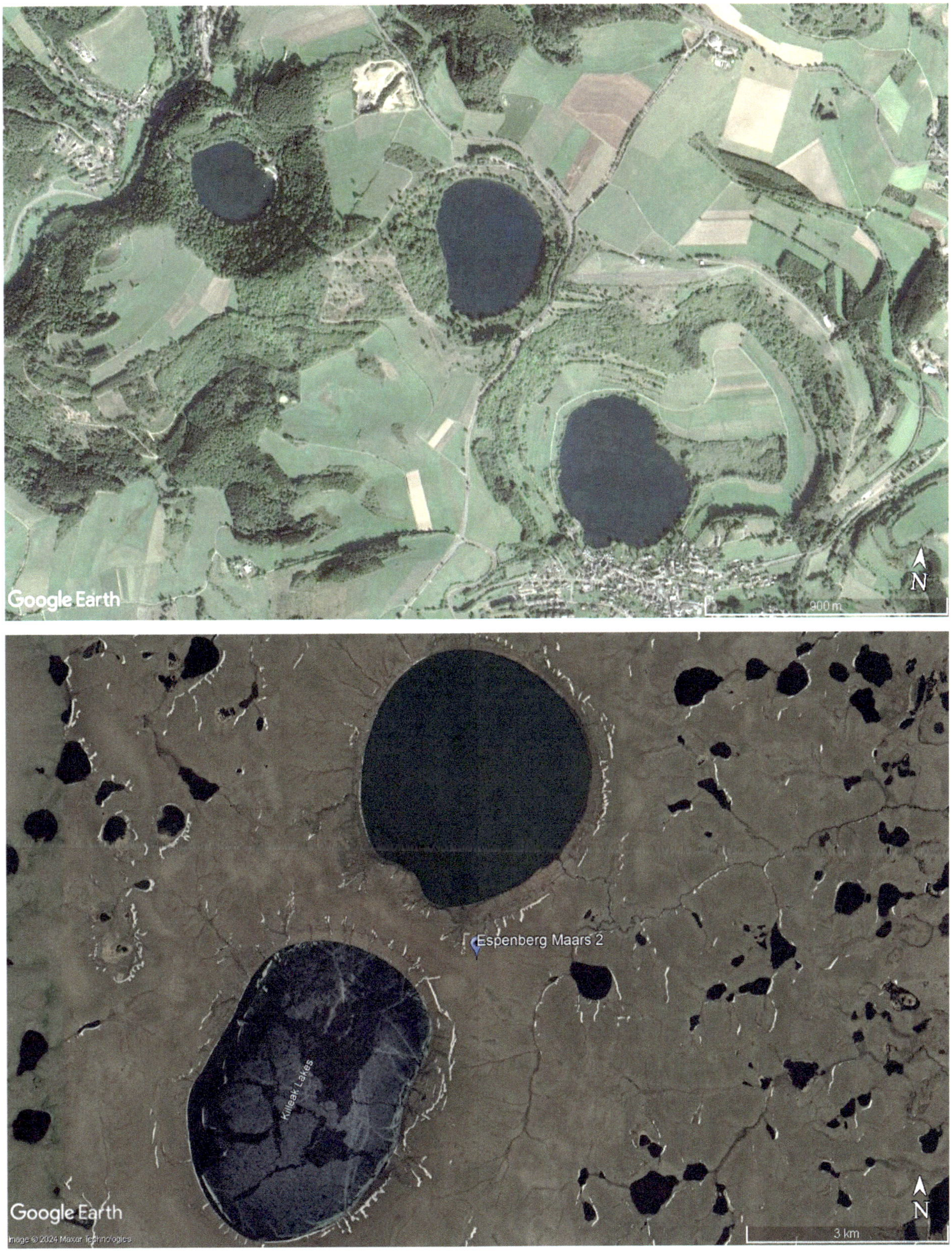

Fig. 4.31 **a** Maars, Germany. 50° 10′ 27.22″ N, 6° 50′ 59.07″ E. ©Google Earth. Scale bar = 900 m. **b** Maars, Espenberg, Alaska, USA. 66° 20′ 39.78″ N, 164° 3′ 15.38″ W. ©Google Earth. Scale bar = 3 km

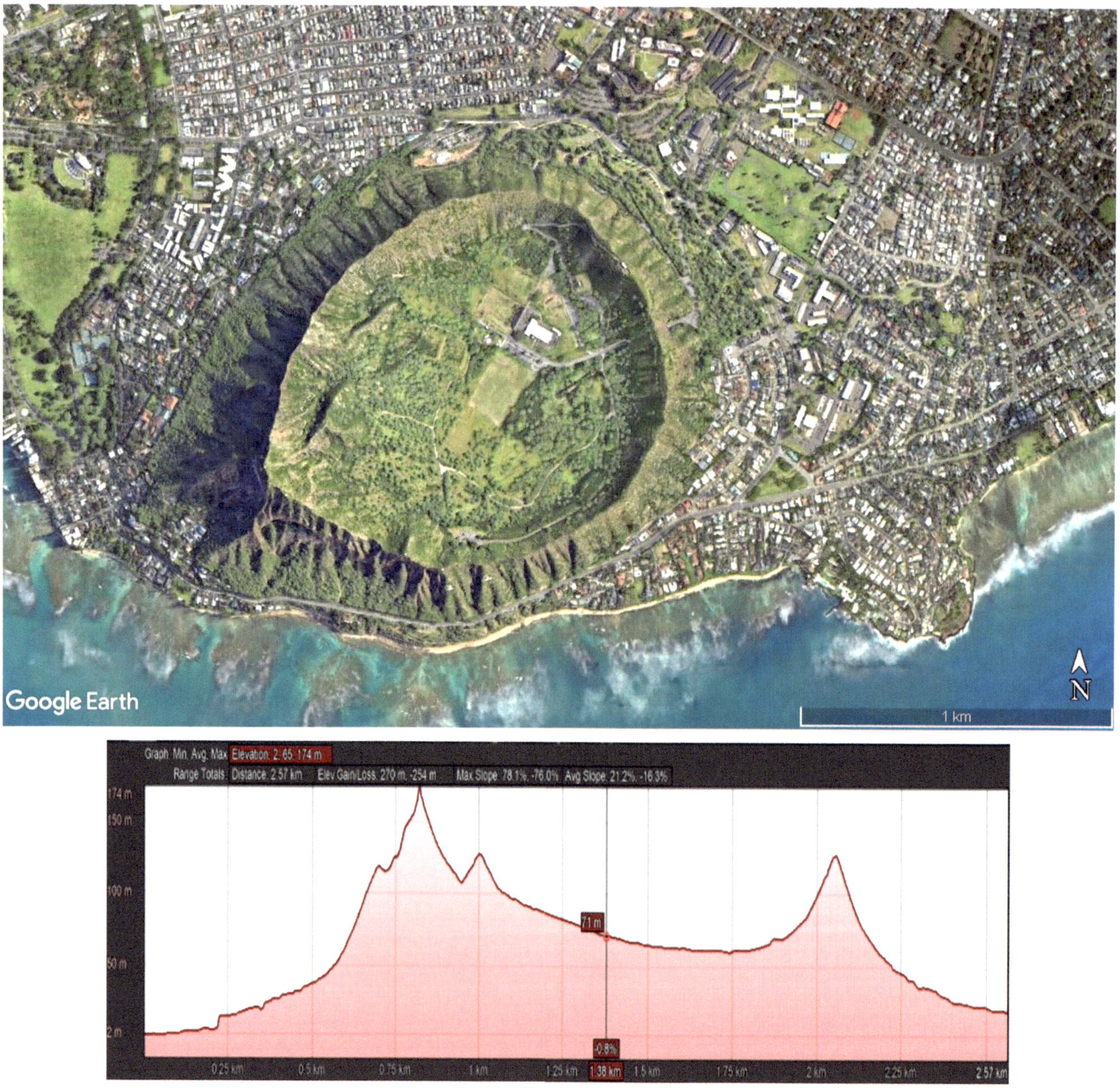

Fig. 4.32 Diamond Head tuff cone, Oahu, USA. 21° 15′ 43.48″ N, 157° 48′ 21.71″ W. ©Google Earth. Scale bar = 1 km. Lower diagram shows a west to east elevation cross-profile

Ḥarrat Khaybar is a volcanic area located north of Medina in Saudi Arabia. It covers c 12,000 km^2 and contains a wide range of volcanic lithologies and spectacular landforms. The most recent eruption occurred between 600 and 700 AD. Its lava fields were formed by a series of eruptions over the course of 5 million years. These eruptions took place along a series of vents stretching over a 100 km north–south linear vent system (Alohali et al. 2022). Harrat Khaybar is one of the 180,000 km^2 of Arabian lava fields (*harrah* in Arabic), which form one of the largest basaltic provinces in the world. They span the western side of the Arabian Peninsula from Yemen in the south through Saudi Arabia and Jordan to Syria and Turkey in the north (Edgell 2006, Sect. 14.2.2).

Magmatism in the region is divided into two main phases. The first and older period (c 30 to 20 mya) was associated with passive mantle upwelling, which was contemporaneous with the opening of the Red Sea. This produced eruptions of lavas along a structural trend that lies parallel to the Red Sea. A subsequent period of volcanism (ca. 12 mya to the present) has been attributed to the Afar mantle plume beneath western Arabia developing extensional stresses and uplift.

Jabal Qidr (Fig. 4.33), a stratovolcano, is built from several generations of dark, fluid basalt lava flows, while

(a)

(b)

Fig. 4.33 **a** Harrat Khaybar, Saudi Arabia. 25° 38′ 26.84″ N, 39° 57′ 15.28″ E. ©Google Earth. Scale bar = 8 km. **b** A close-up view of Jabal Bayda tuff ring in Harrat Khaybar, Saudi Arabia. 25° 39′ 39.97″ N 39°56′ 19.68″ E. ©Google Earth. Scale bar = 1 km. It has a small vesicular comendite dome in the crater floor

Jabal Abyad was formed from a more viscous, silica-rich rhyolitic lava. The 322 m high Jabal Qidr exhibits the classic cone shape of a stratovolcano and is believed to have erupted in historic time. It has a c 400 m diameter summit crater. It produced c 18 km long dark pahoehoe lava flows, which mainly flowed southwest, and a large fan of fall-out lapilli that accumulated up to 15 km east of the summit. Jabal Abyad is a lava dome with a diameter of c 600 m and is a rounded mass of thicker, more solidified lava flows. To the west is the impressive light coloured mass of Jabal Bayda. This symmetric structure is a tuff cone, formed by eruption of lava in the presence of water. Locally it is often called 'The White Volcano' (Moutfi and Németh 2014).

4.6 Lava Flows, Pyroclastic Flows, and Lahars

Basic lava, such as basalt, is very fluid. It spreads easily, so creating volcanoes of low gradient and with a convex profile. It forms shields, cones and mounds. On the other hand, andesitic lava is highly viscous, moves sluggishly, and forms thick, steep-sided domed extrusions. High silica content in the andesitic lava contributes to causing this high viscosity and low silica content in the basaltic lava means it is of a lower viscosity. Both lavas are formed at different plate boundaries which is why they have different silica contents: the basaltic lava forms from the hot mantle material at constructive plate boundaries and hotspots, whereas the andesitic lava forms at destructive plate boundaries from melting crust.

In the Tibesti Mountains of the central Sahara, lies the extinct caldera of the Trou au Natron (Roland 1974) and the great dormant Pic Toussidé stratovolcanic peak (3265 m above sea level) (Deniel et al. 2015; Permenter and Oppenheimer 2007). It is possible that this Saharan volcano is caused by continental volcanism over a hotspot tied to a deep-seated mantle plume. On its western flanks there are large, fresh, lava flows extending outwards by as much as c 20 km (Fig. 4.34). The Trou au Natron sits at an elevation of 2450 m and has an irregular diameter of approximately 6–8 km and is up to 1000 m deep. Four smaller volcanic cones, made of scoria or tuff sit on the floor of the caldera, and numerous smaller vents and hot springs on the caldera's floor emit hot steam and mineral-rich water. During the late Pleistocene the caldera was occupied by a lake that may have been up to 500 m deep.

In Arizona, USA, near to Flagstaff, is an impressive example of an a'a lava flow (Fig. 4.35a). It is over 50

Fig. 4.34 Tibesti, Chad, showing Pic Toussidé and its lava flows on its western side, and the Trou au Natron crater in the far south east. 21° 4′ 14.69″ N, 16° 30′ 41.88″ E. ©Google Earth. Scale bar = 10 km

m thick and more than 5.5 km long. An a'a lava flow Is formed by rather viscous flows, and creates a rough surface. By contrast there are smoother lava surfaces called *Pahoehoe*, which is derived from the Hawaiian word for 'ropy'. Called the Black Point Lava Flow, it is part of Arizona's San Francisco Volcano Field (Riggs et al. 2019). The lava flowed eastward over the older Permian and Triassic sedimentary rock sequences (spanning the period from roughly 300 to 240 million years ago) to the Little Colorado River.

Big and Little Glass Mountains are rhyolitic obsidian lava flows, estimated to be between 300 and 1100 years old, on the Medicine Lake Highlands shield volcano in northern California. They are remarkable for their flow structures, which are in many respects similar to those of glaciers (Fink 1980a) (Fig. 4.35b). They have been interpreted as folds (Fink 1980b).

There are many cases where the superior resistance of lava flows can lead, after downward erosion, to the formation of Inverted relief (de Vries et al. 2022). One example of this is a 20 km long ridge in central Turkey (Fig. 4.36). This is composed of a Mid-Pleistocene (1.85 mya) inverted lava flow with a Holocene flow running at the eastern base, south of Tuzköy near Goreme in Türkiye. The younger lava is about 100 m below the top of the ridge.

In the Pacific Ocean off Mexico, there is a small volcanic island called San Benedicto. Here a lava flow has formed a delta in the sea (Fig. 4.37). Flow of magma was from a side fissure which breached a sea-cliff eroded into pyroclastic debris. The Barcena volcano, located on the southern end of the island, was formed by eruptions during 1952 and 1953, with the magma breaking through the cone on December 8th 1952. By late 1953 the volcano was dormant again (Richards and Dietz 1956). Such lava deltas are quite common on the flanks of volcanic islands (Rodriguez-Gonzalez et al. 2022).

In the western Indian Ocean lies La Réunion. Volcanism has built this up from the ocean floor as a 7 km high cone over the last five million years (Firode et al. 2024). Réunion Island has a subaerial extent of around 2500 km^2. The island is a volcanic complex composed of at least three edifices. Two of them are currently visible in the landscape—the Piton des Neiges (3070 m) and the Piton de la Fournaise (2632 m)—and the third one, called Les Alizés, considered to have grown before about 600,000 years ago, is now completely destroyed by large mass-wasting events and erosion. With Réunion's high tropical rainfall, steep slopes and relatively young relief, rates of erosion and sediment transport are huge and often considered some of the highest in the world. One of its volcanoes is among the most active on Earth - Piton de la Fournaise. This is a shield volcano on the island's eastern side. Most eruptions of this volcano have been of the Hawaiian type, with very fluid basaltic lava flowing out, and with fire fountaining at the vent. In April 2007, the volcano erupted and over a period of just 30 days produced an estimated 3,000,000 m^3 of lava per day. Much of this flowed eastwards down into the Indian Ocean (Bachèlery et al. 2010) and was associated with the collapse of the summit's Dolomieu Crater (Fig. 4.38).

Mount St. Helens is an active stratovolcano located in Washington State, USA (Tilling et al. 1990). It is geologically young compared with the other major volcanoes in the Cascade Mountains. It has formed only within the past 40,000 years, and the summit cone that was present before its 1980 eruption began rising about 2200 years ago. Mount St. Helens is part of the Cascades Volcanic Province. This is an arc-shaped band extending from southwestern British Columbia in Canada through to northern California, roughly parallel to the Pacific coastline. Beneath the Cascade Volcanic Province, a dense oceanic plate sinks beneath the North American Plate via subduction.

The Mount St. Helens major eruption, which took place in 1980, remains the deadliest and most economically destructive volcanic event in U.S. history (Fig. 4.39). A massive debris avalanche, triggered by a magnitude 5.1 earthquake, caused a lateral eruption that reduced the elevation of the mountain's summit from 2950 to 2549 m (Gabrielli et al. 2020). This left a 1.6 km wide horseshoe-shaped crater. By the end of April, the north side of the mountain had started to bulge, and on May 18th a second earthquake, also of magnitude 5.1, triggered a massive collapse of the north face of the mountain. Magma welled out, creating a large-scale pyroclastic flow. The collapse material became mixed with ice, snow, and water to create lahars. These flowed many km down the Toutle and Cowlitz rivers, causing much damage.

As the case of Mt St. Helens shows, volcanoes can collapse. One of the biggest cases of this is at Socompa in northern Chile (Francis et al. 1985) (Fig. 4.40a). At some point between 10,000 and 500 years ago a major debris avalanche travelled 35 km in a north westerly direction and was followed by another that travelled more than 40 km. Some of the bocks forming the avalanche deposits are gigantic—3 km long and 50 m high. Another instance of a debris avalanche coming from a volcano is in neighbouring Bolivia at Tata Sabaya (Fig. 4.40b). Following construction of the stratovolcano, the edifice collapsed catastrophically resulting in the large butterfly-shaped debris apron which extends southwards of the volcano into the Salar de Coipasa. The debris avalanche deposit covered ~300 km^2 of this salt lake. The collapse scarp is well preserved, defining an amphitheatre which accounts for almost a 120° sector of the volcano (Francis and Wells 1988; da Silva et al. 1993). The debris hummocks vary in size from over a hundred meters in height at the proximal end to a few meters at its exposed distal portions. The large mega-blocks close to

(a)

(b)

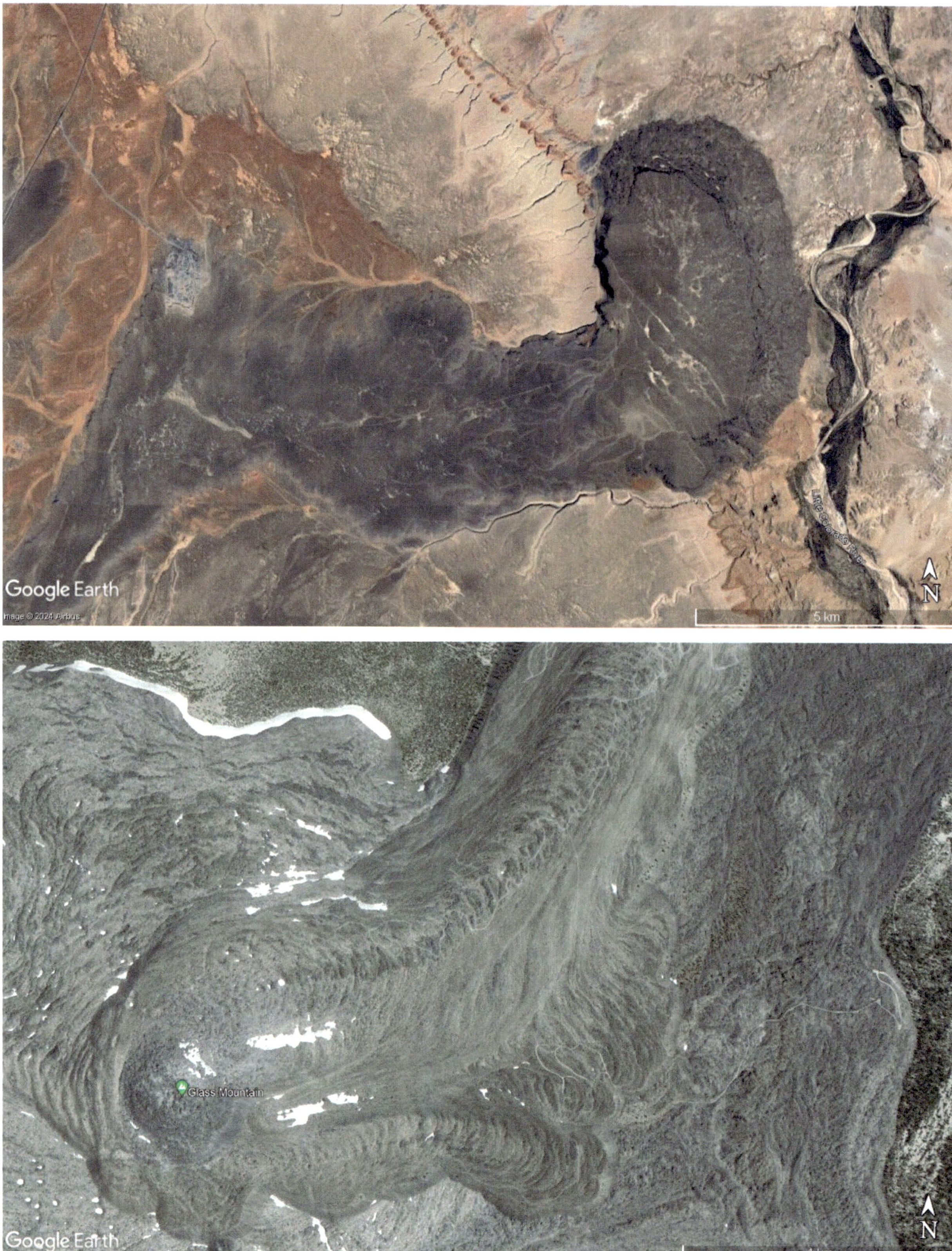

Fig. 4.35 **a** Black Point lava flow near Flagstaff in Arizona, USA. 35° 40′ 42.32″ N, 111° 23′ 23.69″ W. ©Google Earth. Scale bar = 5 km. The Little Colorado River is in the east of the image. **b** Obsidian lava flows on Glass Mountain, northern California. 41° 36′ 4.58″ N, 121° 29′ 30.37″ W. ©Google Earth. Scale bar = 1 km

Fig. 4.36 Inverted relief associated with a lava flow in central Türkiye. 38° 43′ 12.60″ N, 34° 31′ 19.90″ E. ©Google Earth. Scale bar = 4 km

the volcano consist of large chunks of the upper part of the edifice with some of the original stratigraphy still preserved.

In the far east of Russia lies the Kamchatka Peninsula (Sinnyoysky et al. 2023). This is located along the northwestern border of the subducting Pacific plate and is one of the most active volcanic and seismic regions of the world. It is a World Heritage Site. Kamchatka hosts about 30 active volcanoes. Holocene volcanism has been highly explosive, and numerous tephra horizons mantle the topography. The dominantly andesitic Shiveluch volcano (Fig. 4.41) is one of the most voluminous explosive centers of Kamchatka, with a magma discharge of about 36×10^6 t per year. It rises c 3200 m above its surroundings. Shiveluch activity during the Holocene has been characterized by plinian eruptions alternating with periods of dome growth and subsequent debris avalanches. During that period there have been at least 8 large-scale failures (Belousov et al. 1999). The latest of these was a major plinian eruption in 1964, which was preceded by failure of its southern sector. The area covered by the 1964 debris avalanche deposits was about 98 km^2, with a volume of c 1.5 km^3, and a travel distance of 16 km. A part of the material was displaced as three large toreva blocks which form up to 1.5 km long and 0.15 km high 'stairs' immediately south of the collapse crater (Ponomareva et al. 2006). A toreva block is a single large block of a cliff or escarpment that has slid down more or less intact. There have been subsequent eruptions, and major ash flows occurred in 2005 and 2010.

Oceanic volcanic islands, such as the Canaries, have been especially prone to collapse episodes. The evolution of such islands is generally marked by periods of rapid volcanic construction alternating with episodes of giant sector collapses. Such destabilization events can remove a huge part of the eruptive complexes, either as progressively-creeping rotational slides or in the form of fast running debris avalanches produced by the sudden failure of a flank of the volcanic edifice. The rapid transfer of the landslide products into the ocean can generate large tsunamis (Boulesteix et al. 2012).

Ignimbrites are derived from the emplacement, cooling and welding of pyroclastic flows. The types of landforms associated with ignimbrites depends on the extent to which they are welded. Welded ignimbrites, also called welded tuffs, are usually characterized by extremely low erodibility, often forming plateaux that exist for millions of years, sometimes as inverted relief (Biró et al. 2022). This has been discussed in detail in the context of Argentina by Aguilera et al. (2017) and of the Basin and Range Province in Arizona by Dorn (2023). Ignimbrite landforms include bornhardts or domed inselbergs, talus flatirons, tafoni, case

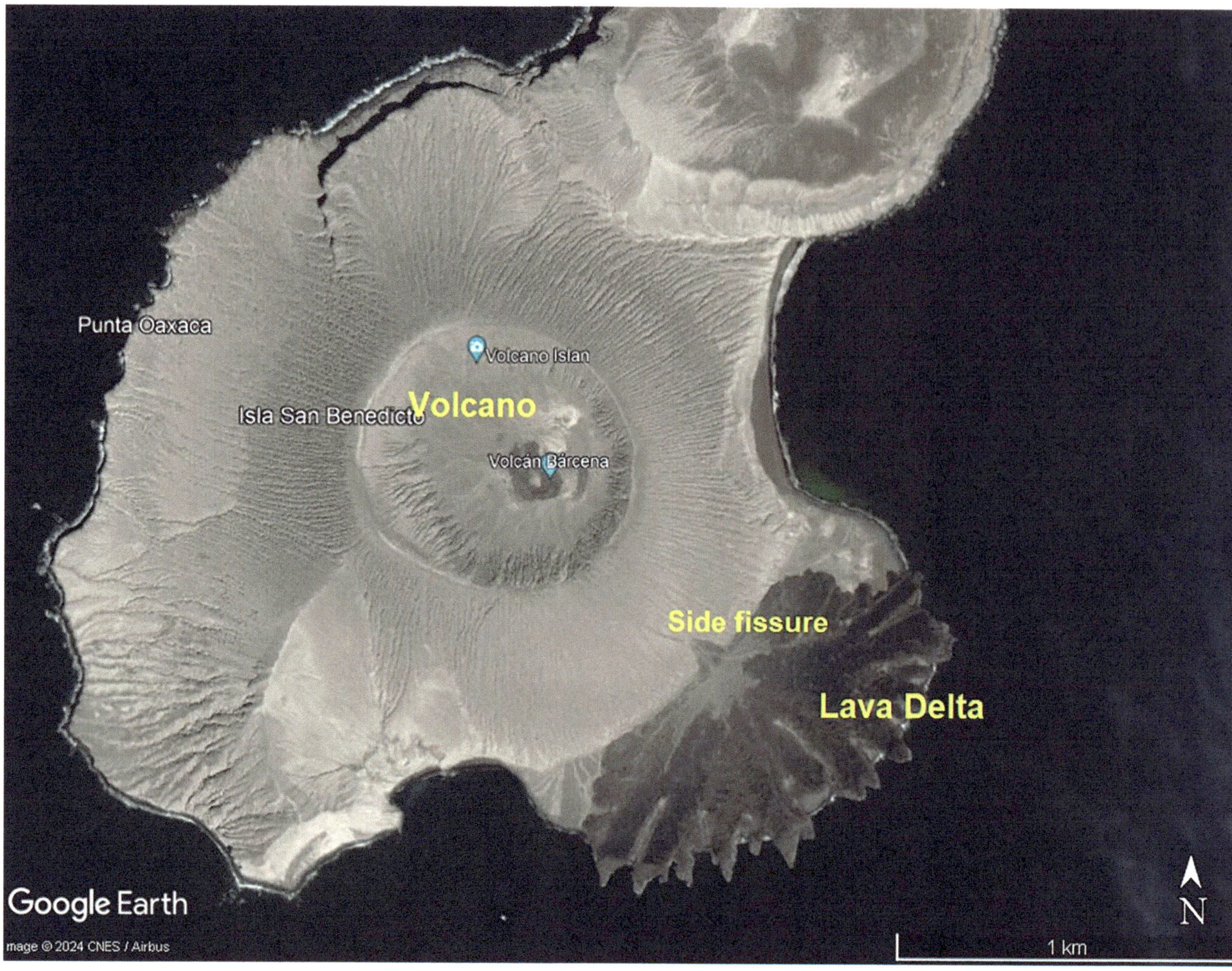

Fig. 4.37 A lava delta formed on San Benedicto Island, Mexico. 19° 18′ 28.58″ N, 110° 48′ 15.67″ W. ©Google Earth. Scale bar = 1 km

hardening, and cliff faces, pediments and columnar forms such as hoodoos or fairy chimneys. Famous examples of the last of these occur in Cappadocia, Türkiye (Sarikaya et al. 2015) (Fig. 4.42). The formation of this landscape dates back to the late Miocene epoch (c 10 mya) when ignimbrites and pyroclastic deposits started to spread out from a few volcanic centres over an area of 20,000 km^2. The evolution of the Cappadocian landscape was governed by the uplift of the area since the late Miocene. Gently sloping plateaux formed by the surface of the ignimbrites were later dissected, usually along fractures of softer unwelded materials, to form mushroom-like, cone-shaped structures known locally as 'fairy chimneys'. They have formed where harder materials are underlain by softer ones (Çiner and Aydar 2019). Typical fairy chimney forms include pillars, columns, towers, obelisks and needles that can reach 50 m in height.

Ignimbrites can also be eroded by wind. This can be seen in the central high Andes of South America (Inbar and Risso 2001). This is a very dry area and fierce winds blow from the northwest for much of the year, furrowing the landscape into parallel *yardangs* (Fig. 4.43). Andean ignimbrites are silicic, and so they contain many quartz phenocrysts. On erosion, therefore, they liberate the agents of their own destruction: quartz grains freed from their matrix and picked up by the wind shot blast the remaining rock (Francis and Oppenheimer 2004, p. 346).

Finally, lahars are mudflows that have developed on the flanks of volcanoes (Vallance 2024). These hazardous phenomena range in size from small volume events (< 0.1 million m^3) to large scale collapses of material (> 3500 million m^3). They are capable of travelling at speeds of up to 150 km per hour and may travel for hundreds of km down valleys (Thouret et al. 2020).

In 1991, as we saw in Fig. 4.12, Mount Pinatubo in the Philippines, suffered a great eruption, and simultaneously Typhoon Yunya deposited a considerable amount of rainfall which led to pyroclastic materials being mobilized and transported down along valleys (Fig. 4.45). Other huge lahars radiate down from the Volcán de Fuego stratovolcano

Fig. 4.38 Couleé de Lave 2007, La Réunion. 21° 14′ 26.68″ S, 55° 45′ 40.06″ E. ©Google Earth. Scale bar = 6 km. The Dolomieu volcano is named after the great French volcanologist, Déodat de Dolomieu (1750–1801). Dolomite was also named after him

in Guatemala, with major flows taking place in 2018 (Cando-Jácome and Martínez-Graña 2019) (Fig. 4.44). They are frequently triggered by intense thunderstorms and rain events between June and October. One of the most disastrous events in recent decades was when pyroclastic material and melted snow produced lahars from the Nevado del Ruiz in Colombia's Central Cordillera. These swept down the Rio Lagunillas, at mean velocities at peak flow of up to 17 m per second, to entomb 23,000 people (three quarters of the population) in Armero (Pierson et al. 1990). Within four hours of the beginning of the eruption, the flows had descended 5100 m and traveled up to 104 km.

4.7 Lava Plateaux, Passive Margins and Great Escarpments

Not all lava flows from central vents. It can also flow from linear fractures, as in south west Iceland. Here basalt flows out along fractures and cones, created by opening up of the mid-oceanic rift as the North American and Euro-Asian plates diverge (Fig. 4.46). The crust in this Eastern Volcanic Zone is cyrrently spreading at c 19 mm per year (Jónsson et al. 1997). An eruption at Lakagigar in 1783 was calamitous, and eruptions are continuing at the present time (Metrich et al. 1991). This type of eruption is called *Icelandic*. Indeed, fissure vents are very widespread in Iceland (Hjartardóttir and Einarsson 2015). In the Palaeocene and early Eocene Britain too was astride this mid-oceanic rift, and basaltic lavas spewed out to give plateaux in locations like Mull, Skye and Antrim (Hansen et al. 2009).

The continental crust in the Pacific Northwest of the USA, Siberia, parts of north western Scotland, southern Brazil, Ethiopia, and Namibia have in times past split widely enough and long enough to allow basaltic lava to flood vast areas of terrain. In the case of Ethiopia, fissures akin to those in Iceland have developed (Wadge et al. 2016). The Snake-Columbia flood basalts in the USA cover over 600,000 km^2 of country and to an average depth of 0.5 km.

The Deccan basalts (traps) of peninsular India cover upto 700,000 km^2 and are bounded on their western side by the great escarpment of the Western Ghats (Jay and Widdowson 2008; Kale et al. 2020) (Fig. 4.47). The Western Ghat escarpment was developed from rifing and separation of India from the Seychelles. It is one of the largest escarpments on a volcanic continental margin in the

Fig. 4.39 Mount St. Helens, USA, pyroclastic flow structures. 46° 8′ 12.08″ N, 121° 53′ 0.43″ W. ©Google Earth. Scale bar = 2 km

world with over 1500 km strike length paralleling the coast and elevations ranging beyond 1 km. The Deccan basalts are an example of a Large Igneous Province (LIP). These represent the most extensive volcanic events on Earth during which massive amounts of melts (> 10^6 km^3) erupt in relatively short periods of time (< 1 Ma). Many of these volcanic provinces were emplaced at the former positions of currently active hot spots (Hoyer et al. 2023). The Deccan lavas were rapidly erupted over a period of 0.5–5 Ma at, or near to, the Cretaceous-Tertiary (K/T) boundary and thicknesses of >1200–1700 m are exposed along the Western Ghats (Widdowson 1997).

One interesting feature of the Panchgani area near Mahableshwar is that the basalts have been deeply weathered to produce extensive spreads of laterite (ferricrete) (Kale 2014) (Fig. 4.48). The Deccan high-level laterites typically form a distinctive 10–20 m high cliff around the margins of the basalt mesas. Laterites and ferricretes are horizons, at or near the land surface, made up of the cementation of surface materials by iron oxides that typically form a resistant duricrust. They are perhaps the most widespread of the common duricrust materials (See Sect. 5.6). The highly chemically and physically durable nature of ferruginous duricrust material has meant that it has often played a prominent role in the development and evolution of tropical, and sub-tropical landscapes. In its broadest sense the terms laterite and ferricrete can be used to describe any duricrust material in which the dominant bulk components are iron-rich compounds (typically iron oxides and sesquioxides). Laterites are Fe-rich duricrusts which have formed directly from the breakdown of materials in their immediate vicinity through residual enrichment during in-situ weathering and the preferential leaching and removal of the more mobile elements and other primary constituents during the weathering process (Widdowson and Gunnell 1999). The Panchgani laterites have formed partly in response to the high available relief and heavy monsoonal rainfall. There has, however, been some discussion as to their precise origin. They have generally been interpreted as the products of in situ weathering but Ollier and Sheth (2008) have raised the possibility that they formed as a result of inversion of relief. This hypothesis is that the laterite caps forming the present summits of the duricrusted mesas and tablelands were formed in a river valley system, and these original valleys have since been inverted as the surrounding, softer weathered basalts have been eroded to a greater extent.

In Namibia there is a large basalt plateau, covering c 78,000 km^2, called Etendeka (Milner et al. 1995; Renne et al. 1996) (Fig. 4.49). It was created by the spewing out of flood basalts as the break-up of Gondwanaland between

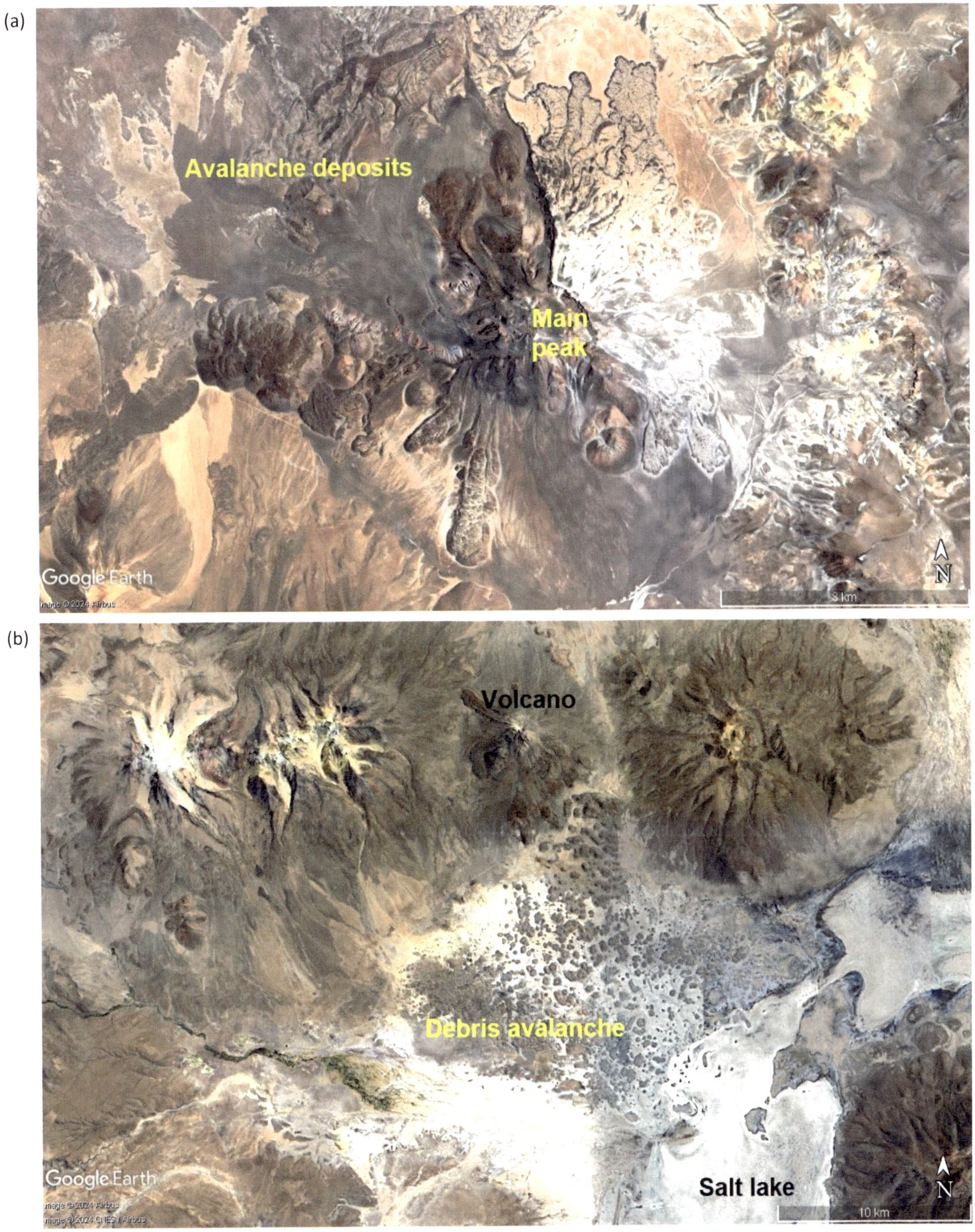

Fig. 4.40 **a** Socompa, Chile, showing avalanche deposits. 24° 23′ 16.71″ S, 68° 15′ 13.62″ W. ©Google Earth. Scale bar = 8 km. **b** Tata Sabaya debris avalanche, Bolivia. 19° 13′ 26.27″ S, 68° 26′ 46.73″ W. ©Google Earth. Scale bar = 10 km

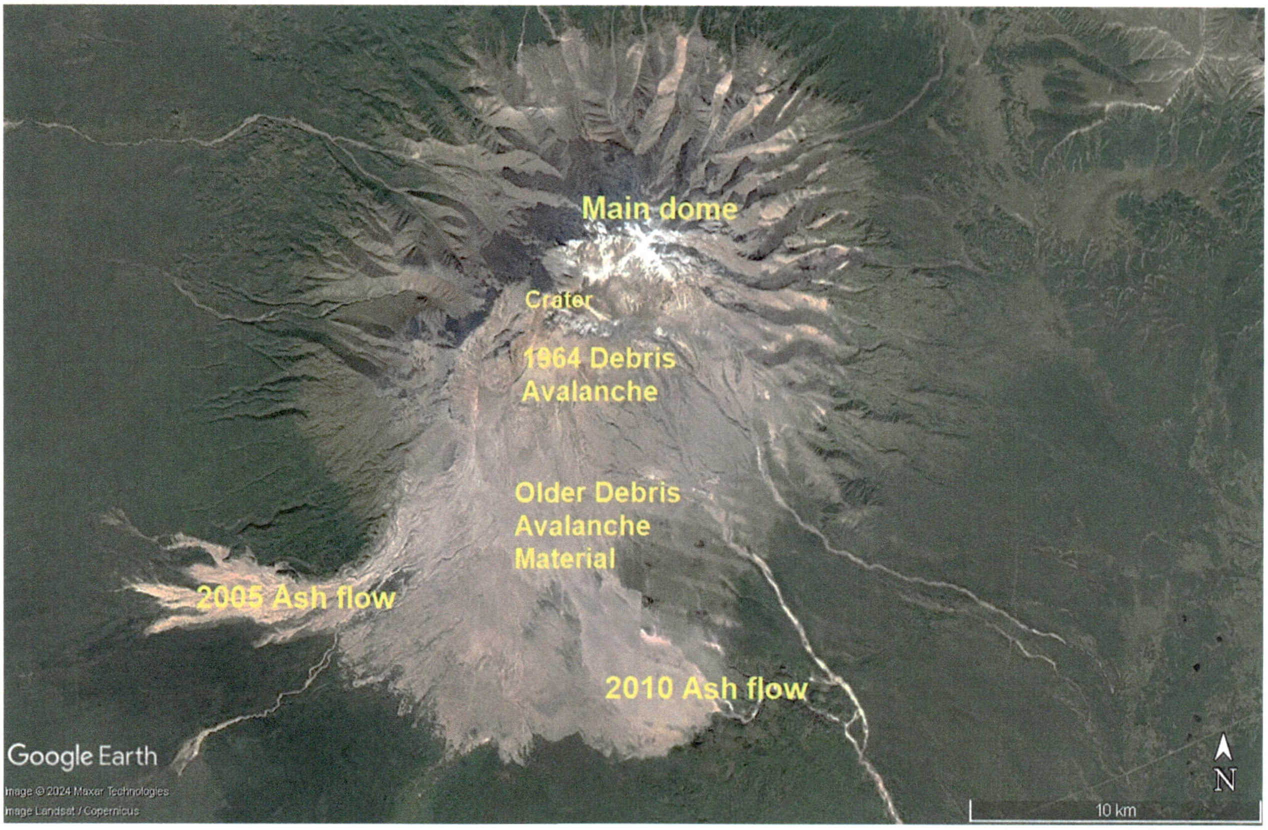

Fig. 4.41 Shiveluch volcano, Kamchatka, Russia. 56° 36′ 35.30″ N, 161° 21′ 1.57″ E. ©Google Earth. Scale bar = 10 km

Fig. 4.42 Ignimbrite landforms: Towers and fairy chimneys at Cappadocia, Türkiye. 38° 39′ 29.69″ N, 34° 51′ 26.22″ E. ©Google Earth. Scale bar = 70 m

Fig. 4.43 Ignimbrite landforms: Yardangs developed in ignimbrites to the east of the Andes in Northwest Argentina at c 25 36′ S, 66 48′ W. ©Google Earth. Scale bar = 4 km

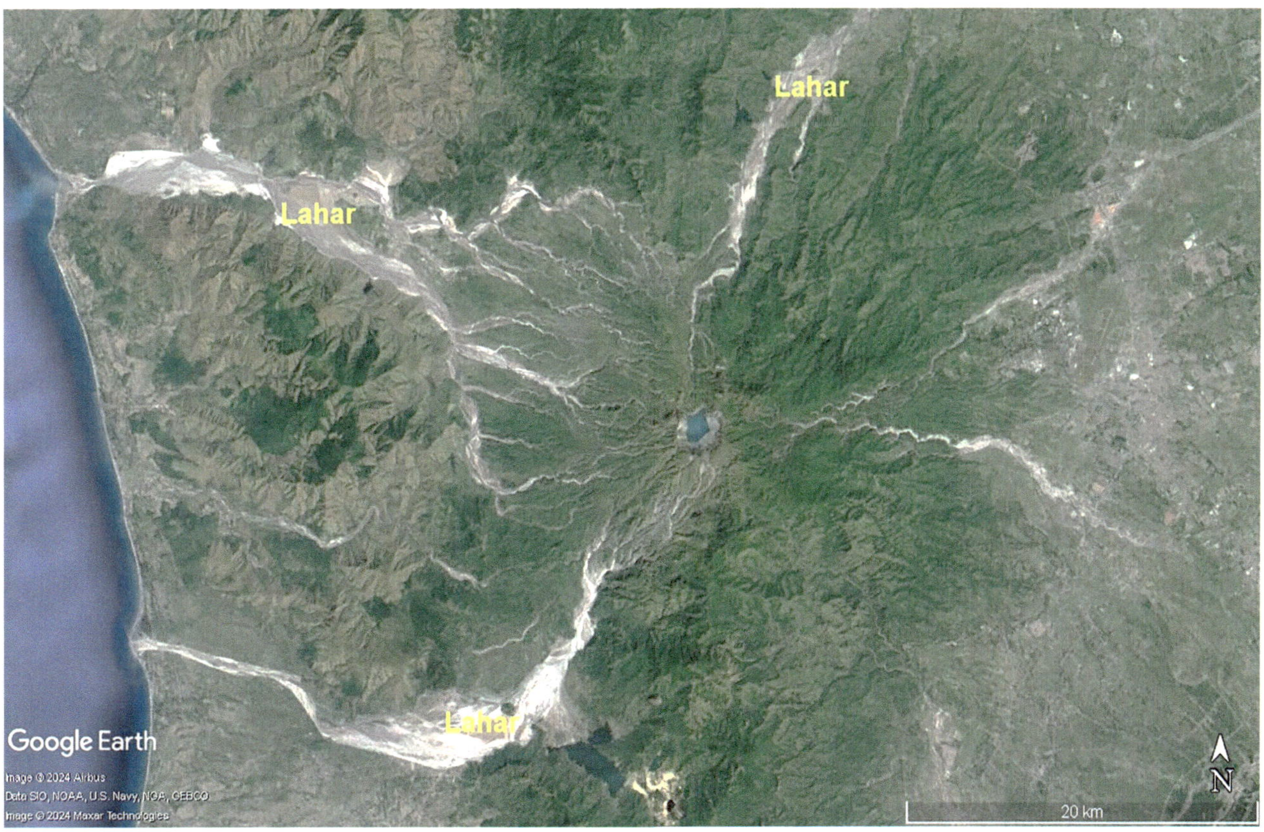

Fig. 4.44 Lahar field, Mount Pinatubo, Phillipines. 15° 9′ 41.14″ N, 120° 22′ 29.86″ E. ©Google Earth. Scale bar = 20 km

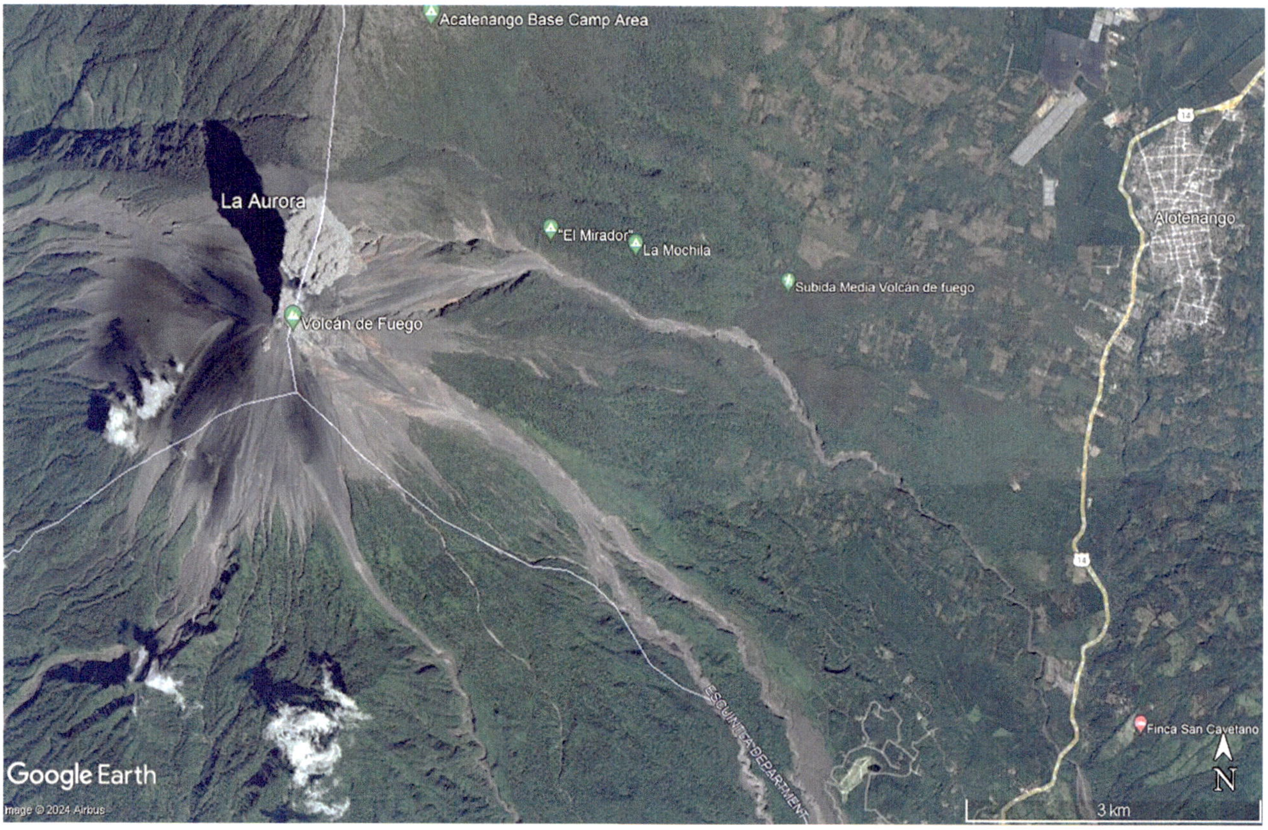

Fig. 4.45 Lahar field, Volcán de Fuego, Guatemala. 14° 28′ 5.51″ N, 90° 51′ 19.13″ W. ©Google Earth. Scale bar = 3 km. The image from January 2024 also shows the volcano erupting

Fig. 4.46 Lakagigar fissure vents, Iceland. 64° 4′ 59.66″ N, 18° 23′ 48.03″ W. ©Google Earth. Scale bar = 6 km

Fig. 4.47 **a** Deccan Plateau, India. 17° 13′ 24.02″ N, 74° 7′ 20.54″ E. ©Google Earth. Scale bar = 90 km. **b** Western Ghats, India, near Panchgani, showing the layering and great thickness of the Deccan Plateau Basalts (*Photo* A. S. Goudie)

Fig. 4.48 Panchgani laterite crust and tablelands, India (*Photo* A. S. Goudie)

South America and Africa occurred. In South America they are called the Paraná Basalts. The Etendeka Plateau towers by 700–800 m above the surrounding plains (which are developed in ancient metamorphic rocks). Layers of different hardness within the lavas produce steps, and valleys are deeply incised. The break-up of this portion of Gondwana and the development of the Atlantic Ocean took place in late Jurassic to early Cretaceous times (c 135 mya).

Comparably aged basalts occur on the other side of the Atlantic in the Iguazu/Iguaçu region of Brazil/Argentina (Fig. 4.50) and give rise to major waterfalls (Lima and Binda 2013; Stevaux and Latrubesse 2010). The falls have an arcuate head c 2.7 km long and have a stepped form due to differences in the hardness of different basalt layers. Across a width of almost three kilometres the river drops vertically some 80 m in a series of cataracts. It is nearly three times wider than the Niagara Falls and significantly wider than the Victoria Falls. The river, aptly named after the indigenous term for 'great water', forms a large bend in the shape of a horseshoe and constitutes the international border between Argentina and Brazil before it flows into the Parana River less than 25 km downstream.

After a global study of waterfalls, Goudie (2020) suggested that although waterfalls occur on many rock types of moderate to high resistance, they are most common on basalt lavas and flat-lying sandstones (Goudie 2020). The Victoria Falls on the Zambezi (Fig. 4.51) form the world's largest sheet of falling water when in flood, and have a width of 1700 m and a maximum vertical drop of 108 m. They fall over Jurassic (Karoo-age) Batoka basalts dating back to 180 mya. The Zambezi then plunges into gorges cut into the basalts and one can see how, as the falls have worked backward over time, the course of the river has been controlled by a series of zigzag joints and faults as the knickpont has retreated. Headward erosion has incised 20 km of gorges below the Falls in the last 300–250 ka (Moore and Cotterill 2010).

Another large basalt plateau occurs in southern Africa in the shape of the Drakensberg-Maluti Mountains of South Africa and Lesotho (Knight and Grab 2015). This great escarpment rises to over 3500 m above sea level, and is a remnant of a great basalt plain (Fig. 4.52). The basaltic lava is over 1500 m thick and covers the sandstones of the Clarens formation, part of the Karoo Supergroup.

(a)

(b)

Fig. 4.49 **a** Dissected Etendeka lava plateau, Namibia. 19° 53′ 4.37″ S, 14° 9′ 11.01″ E. ©Google Earth. Scale bar = 3 km. **b** Hills capped with Etendeka lavas in Namibia (*Photo* A. S. Goudie)

(a)

(b)

Fig. 4.50 **a** The Iguazu/Iguaçu Falls of South America, formed where the Iguazu river flows over the Paraná Basalts. 25° 41′ 24.77″ S, 54° 26′ 17.70″ W. ©Google Earth. Scale bar = 600 m. **b** The Iguazu/ Iguaçu Falls (Photo Alice Goudie)

(a)

(b)

Fig. 4.51 **a** The Victoria Falls, Zambia/Zimbabwe, showing the Falls and the structurally-controlled zigzag course through the downstream gorges. 17° 55′ 47.92″ S, 25° 51′ 7.93″ E. ©Google Earth. Scale bar = 2 km. **b** The Victoria Falls and gorge from the Zimbabwe side of the Zambezi (*Photo* A. S. Goudie)

It was erupted c 183 mya (i.e. of similar age to the basalts at Victoria Falls) and heralded the break-up of the Gondwanaland supercontinent. The Tugela Falls, the highest in Africa, with a total drop of 947 m, plunge over the basalts and sandstones of the escarpment face.

In Ethiopia, the Simien Mountains, a World Heritage Site, are made up by a large lava plateau (Asrat et al. 2012) that was formed prior to the formation of the Main Ethiopian Rift Valley. The Cenozoic era witnessed the uplifting of the Ethio-Arabian landmass, and this was

(a)

(b)

Fig. 4.52 **a** Cathedral Peak, South Africa. 28° 57′ 42.77″ S, 29° 11′ 22.09″ E. ©Google Earth. Scale bar = 2 km. **b** Drakensberg Mountains, South Africa, near Champagne Castle (*Photo* A. S. Goudie)

followed by deep seated crustal fracturing and subsequent outpouring of huge quantities of lava. This in turn was followed by the breaking up of the uplifted dome and formation of the Rift valley during the Middle Cenozoic. The massif was subsequently dissected by deep erosional gorges formed by the major west or south-westward draining river basins of the Blue Nile, Tekeze, Omo and their tributaries. One of Africa's major highlands, it forms a spectacular monument to the power of erosion and the landscape is characterized by deep valleys, jagged mountains, steep cliffs and precipices of more than 1500 m in height (Fig. 4.53). It contains the highest point in the country, the Ras Dashen, at an altitude of 4624 m a.s.l. The Simien Mountains represent a major Oligo-Miocene volcanic shield built over Oligocene flood basalts, which altogether form the North-western Ethiopian Highlands. The Ethiopian flood basalts (the traps), mostly erupted around 30 Ma ago during a short 1 Ma period. Their thickness reaches up to 2 km in some areas.

The lava plateaus and great escarpments already mentioned in this section are in most cases associated with what are called passive margins—zones where plate movement does not lead to collisions and subduction. Rather, the splitting of Earth's crust leads to eruptions of magma and flexure and uplift of the crust (see Summerfield 1991, pp. 95–102). It is these two processes that lead to the development of great escarpments, such as those of western India, eastern South America, Namibia and eastern South Africa. Another example of a great escarpment associated with great spreads of lava (see also Sect. 4.5) and the opening up of the Red Sea, is that on the western margin of Arabia (Fig. 4.54). This may be Earth's largest. It includes Saudi Arabia's highest peak (Mt. Al Soudah, 3015 m) and Yemen's highest peak (Jabal an-Nabi, 3650 m). It drops more than 1500 m in less than five km distance and its morphology remains almost uniform for about 1000 km, separating rugged terrain on the coastal side below it from the high-elevation Arabian Shield plateau inland. Turab et al. (2023, p. 13) have proposed the escarpment developed in two stages:

> The first stage involved Oligocene and Miocene dome-shaped rift flank uplift, downwarping and additional flexural uplift in response to the formation of oceanic lithosphere in the Red Sea at 13 Ma. This stage involved about 1.0 – 1.5 kilometers of erosion that largely removed the Phanerozoic sediment sequence that had covered the entire Red Sea region at the end of the Mesozoic. The second stage involves the onset of orographic precipitation, substantial erosion of at least two kilometers on the coastal side of the divide, and scarp retreat in the last 5 Ma of the rift evolution.

Fig. 4.53 Simien Mountains, Ethiopia. 13°14′ 21.93″ N, 38° 16′ 0.78″ E. ©Google Earth. Scale bar = 10 km

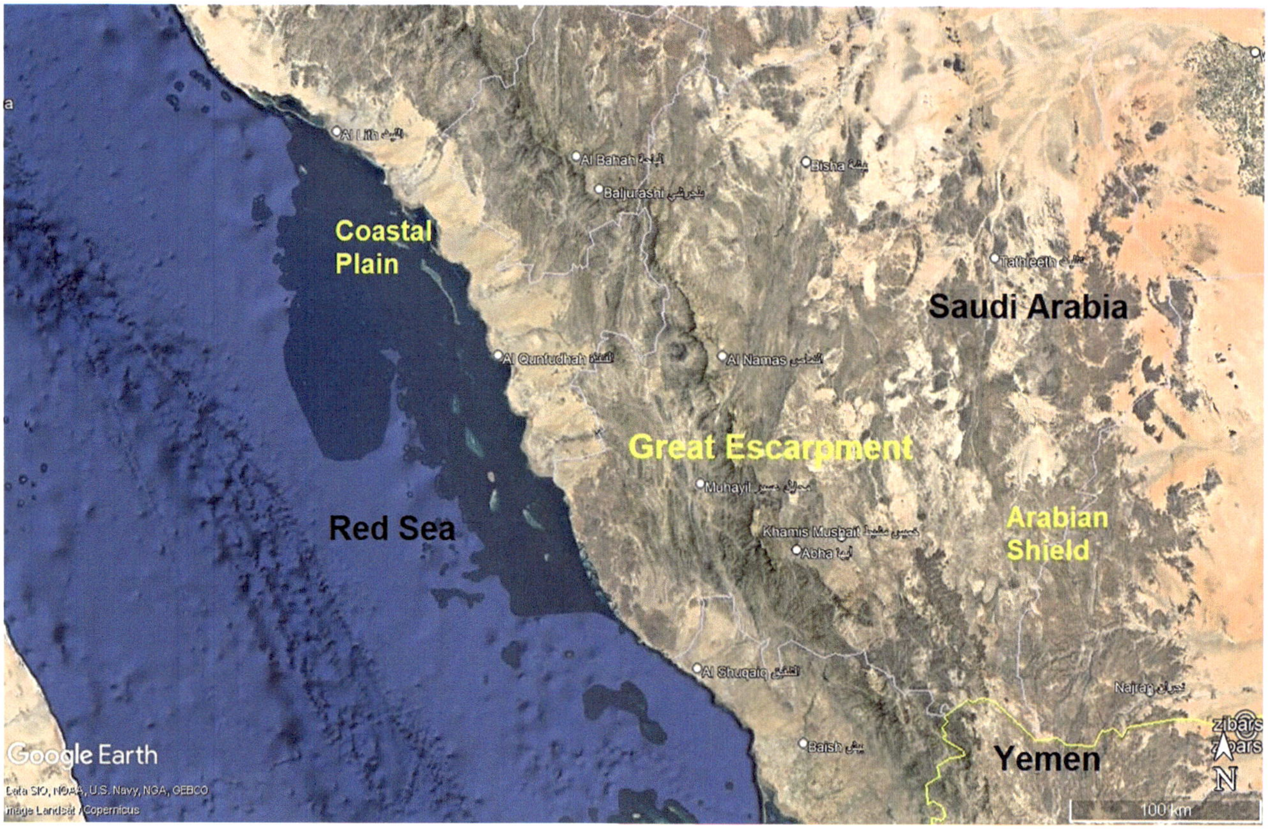

Fig. 4.54 The Great Escarpment of Arabia. 19° 26′ 36.51″ N, 41° 45′ 54.52″ E. ©Google Earth. Scale bar = 100 km

A further study of the evolution of the Arabian margin is provided by Delaunay (2024).

4.8 Conclusion

Volcanoes are impressive landform features that occur in many parts of the world, whether Antarctica or the centre of the Sahara. Most of them are associated with the movement of tectonic plates, and develop through subduction, continental break-up, and movement over mantle plumes (hot spots). In addition to volcanic edifices themselves (e.g. craters, calderas, cones, maars, etc.), volcanic landscapes are also characterised by lava and pyroclastic flows and lahars, and may in some cases be subjected to extreme erosion by marine or subaerial processes.

References

Aguilera EY, Hernando I, Rabassa J (2017) Landscapes developed on ignimbrites. In: Rabassa J (ed) Advances in geomorphology and quaternary studies in Argentina. Springer, Cham, pp 1–48

Alfianti H et al (2022) Bromo activity over the last decade: consistent passive degassing and source magma evolution. Geosci Lett 9(1):15

Allen CD (ed) (2017) Landscapes and landforms of the lesser Antilles. Springer, Cham, p 321

Alohali A et al (2022) Spatio-temporal forecasting of future volcanism at Harrat Khaybar, Saudi Arabia. J Appl Volcanol 11(1):12

Asrat A, Demissie M, Mogessie A (2012) Geoheritage conservation in Ethiopia: the case of the Simien Mountains. Quaest Geogr 31(1):7–23

Atkinson A, Griffin TJ, Stephenson PJ (1975) A major lava tube system from Undara Volcano, North Queensland. Bull Volcanol 39:266–293

Azzouni-Sekkal A, Bonin B, Benhallou A, Yahiaoui R, Liégeois JP (2007) Cenozoic alkaline volcanism of the Atakor massif, Hoggar, Algeria. Geol Soc Am Spec Paper 418:321–340

Bachèlery P et al. (2010) Huge lava flows into the sea and caldera collapse, April 2007, Piton de la Fournaise volcano. Geophys Res Lett 34:L21301

Bacon CR et al (2002) Morphology, volcanism, and mass wasting in Crater Lake, Oregon. Geol Soc Am Bull 114(6):675-692

Bardintzeff JM, Deniel C, Guillou H, Platevoet B, Télouk P, Oun KM (2012) Miocene to recent alkaline volcanism between Al Haruj and Waw an Namous (southern Libya). Int J Earth Sci 101:1047–1063

Begét JE, Hopkins DM, Charron SD (1996) The largest known maars on Earth, Seward Peninsula, northwest Alaska. Arctic 49:62–69

Belousov A, Belousova M, Voight B (1999) Multiple edifice failures, debris avalanches and associated eruptions in the Holocene history of Shiveluch volcano, Kamchatka, Russia. Bull Volcanol 61:324–342

Bertagnini A, Di Roberto A, Pompilio M (2011) Paroxysmal activity at Stromboli: lessons from the past. Bull Volcanol 73:1229–1243

Berthod C et al (2021) The 2018-ongoing Mayotte submarine eruption: magma migration imaged by petrological monitoring. Earth Planet Sci Lett 571:article no. 117085

Biró T, Hencz M, Telbisz T, Cseri Z, Karátson D (2022) The relationship between ignimbrite lithofacies and topography in a foothill setting formed on Miocene pyroclastics–a case study from the Bükkalja, Northern Hungary. Hungarian Geogr Bull 71(3):213–229

Boulesteix T, Hildenbrand A, Gillot PY, Soler V (2012) Eruptive response of oceanic islands to giant landslides: new insights from the geomorphologic evolution of the Teide-Pico Viejo volcanic complex (Tenerife, Canary). Geomorphology 138(1):61–73

Branca S, Chester D, De Beni E, Duncan A (2017) Landforms and landscapes of Mount Etna (Sicily): relationships between a volcano, its environment and human activity. In: Soldati M, Marchetti M (eds) Landscapes and landforms of Italy. Springer, Cham, pp 476–478

Bullard FM (1962) Volcanoes. Univ Texas Press, Austin, p 441

Cando-Jácome M, Martínez-Graña A (2019) Determination of primary and secondary lahar flow paths of the Fuego volcano (Guatemala) using morphometric parameters. Remote Sens 11(6):727

Carracedo JC et al (2007) Eruptive and structural history of Teide Volcano and rift zones of Tenerife, Canary Islands. Geol Soc Am Bull 119(9–10):1027–1051

Chesner CA (2012) The Toba caldera complex. Quat Int 258:5–18

Chester DK, Duncan AM, Guest JE, Kilburn C (2012) Mount Etna: the anatomy of a volcano. Stanford University Press, Stanford, p 412

Çiner A, Aydar E (2019) A fascinating gift from volcanoes: the fairy chimneys and underground cities of Cappadocia. In: Kuzucuoğlu C, Çiner A, Kazancı N (eds) Landscapes and landforms of Turkey. Springer, Cham, pp 534–549

Criado C, Paris R (2005) Volcanic landscape and geomorphological evolution of Tenerife Islands. In: Desir G, Gutiérrez F, Gutiérrez M (eds) Sixth international conference on geomorphology, field trip guide, 2, Zaragoza, pp 473–504

Darwin CR (1842) The structure and distribution of coral reefs. Smith Elder, London, p 214

Delaunay A, Baby G, Paredes EG, Fedorik J, Afifi AM (2024) Evolution of the Eastern Red Sea Rifted margin: morphology, uplift processes and source-to-sink dynamics. Earth Sci Rev 28:104698

Deniel C, Vincent PM, Beauvilain A, Gourgaud A (2015) The Cenozoic volcanic province of Tibesti (Sahara of Chad): major units, chronology, and structural features. Bull Vulcanol 77:1–21

de Silva SL, Davidson JP, Croudace IW, Escobar A (1993) Volcanological and petrological evolution of volcano Tata Sabaya, SW Bolivia. J Volcanol Geotherm Res 55(3–4):305–335

de Vries BV, Karatson D, Gouard C, Németh K, Rapprich V, Aydar E (2022) Inverted volcanic relief: Its importance in illustrating geological change and its geoheritage potential. Int J Geoheritage Parks 10(1):47–83

Dorn RI (2023) Decay of ignimbrite fairy chimneys of Arizona's Basin and Range Province, USA. Earth Surf Proc Landf 48(12):2349–2367

Druitt TH et al (1999) Santorini volcano. Geol Soc Lond Mem 19:165

Edgell HS (2006) Arabian deserts. Nature, origin and evolution. Springer, Dordrecht, p 592

Fink JH (1980a) Gravity instability in the Holocene Big and Little Glass Mountain rhyolitic obsidian flows, northern California. Tectonophysics 66(1–3):147–166

Fink JH (1980b) Surface folding and viscosity of rhyolite flows. Geology 8(5):250–254

Firode L, Duputel Z, Ferrazzini V, Lengliné O (2024) Seismicity under a dormant volcano: unveiling active crustal faulting beneath Piton des Neiges, La Réunion. Bull Seismol Soc Am 114(3):1626–1638

Folkes CB, Wright HM, Cas RA, de Silva SL, Lesti C, Viramonte JG (2011) A re-appraisal of the stratigraphy and volcanology of the Cerro Galán volcanic system, NW Argentina. Bull Volcan 73:1427–1454

Ford RJ (2023) Volcano. Live, dormant and extinct volcanoes around the world. Amber Books, London, p 223

Francis P, Oppenheimer C (2004) Volcanoes (2nd edn). Oxford Univ Press, Oxford, p 521

Francis P, Wells GL (1988) Landsat Thematic Mapper observations of debris avalanche deposits in the Central Andes. Bull Volcanol 50:258–278

Francis PW, Gardeweg M, Ramirez CF, Rothery DA (1985) Catastrophic debris avalanche deposit of Socompa volcano, northern Chile. Geology 13(9):600–603

Friedrich WL, Kromer B, Friedrich M, Heinemeier J, Pfeiffer T, Talamo S (2006) Santorini eruption radiocarbon dated to 1627–1600 BC. Science 312:548–548

Gabrielli S, Spagnolo M, De Siena L (2020) Geomorphology and surface geology of Mount St. Helens volcano. J Maps 16(2):585–594

Giordano G, Cas R, Wright JV (eds) (2024) The geology of volcanoes and their facies models: morphology, dynamics, evolution, successions, economic significance and hazards. Springer, Cham, p 1833

Gischler E et al (2016) Late Quaternary barrier and fringing reef development of Bora Bora (Society Islands, south Pacific): first subsurface data from the Darwin-type barrier-reef system. Sedimentology 63(6):1522–1549

Goudie AS (2020) Waterfalls: forms, distribution, processes and rates of recession. Quaest Geogr 39(1):59–77

Grove AT, Street FA, Goudie AS (1975) Former lake levels and climatic change in the Rift Valley of southern Ethiopia. Geogr J 141:177–194

Guilcher A (1988) Coral reef geomorphology. Wiley, Chichester, p 228

Hampton SJ, Cole JW (2009) Lyttelton Volcano, Banks Peninsula, New Zealand: primary volcanic landforms and eruptive centre identification. Geomorphology 104(3–4):284–298

Hansen J, Jerram DA, McCaffrey K, Passey SR (2009) The onset of the North Atlantic igneous province in a rifting perspective. Geol Mag 146:309–325

Hoyer PA, Haase KM, Regelous M, Fluteau F (2023) Systematic and temporal geochemical changes in the upper Deccan lavas: implications for the magma plumbing system of flood basalt provinces. Geochem, Geophys, Geosyst 24(2):e2022GC010750

Inbar M, Risso C (2001) Holocene yardangs in volcanic terrains in the southern Andes, Argentina. Earth Surf Proc Landf 26(6):657–666

Jay AE, Widdowson M (2008) Stratigraphy, structure and volcanology of the SE Deccan continental flood basalt province: implications for eruptive extent and volumes. J Geol Soc 165(1):177–188

Jónsson S, Einarsson P, Sigmundsson F (1997) Extension across a divergent plate boundary, the Eastern Volcanic Rift Zone, south Iceland, 1967–1994, observed with GPS and electronic distance measurements. J Geophys Res: Solid Earth 102(B6):11913–11929

Hitchcock CH (1906) Geology of Diamond Head, Oahu. Geol Soc Am Bull 17(1):469–484

Hjartardóttir ÁR, Einarsson P (2015) The interaction of fissure swarms and monogenetic lava shields in the rift zones of Iceland. J Volcanol Geothermal Res 299:91–102

Kale VS (2014) The laterite-capped Panchgani tableland, Deccan traps. In: Kale VS (ed) Landscapes and landforms of India. Springer, Dordrecht, pp 217–222

Kale VS, Chatterjee P, Pande K (2020) Emplacement history and evolution of the Deccan Volcanic Province. India. Episodes 43(1):278–299

Kinnison DE, Grant KE, Connell PS, Rotman DA, Wuebbles DJ (1994) The chemical and radiative effects of the Mount Pinatubo eruption. J Geophys Res: Atmos 99(D12):25705–25731

Knight J, Grab S (2015) The Drakensberg escarpment: mountain processes at the edge. In: Grab S, Knight J (eds) Landscapes and landforms of South Africa. Springer, Cham, pp 47–55

Leonard GS et al (2021) Ruapehu and Tongariro stratovolcanoes: a review of current understanding. NZ J Geol Geophys 64(2–3):389–420
Le Turdu C et al (1999) The Ziway-Shala lake basin system, Main Ethiopian Rift: influence of volcanism, tectonics, and climatic forcing on basin formation and sedimentation. Palaeogeogr Palaeoclim Palaeoecol 150(3–4):135–177
Lima AG, Binda AL (2013) Lithologic and structural controls on fluvial knickzones in basalts of the Parana Basin, Brazil. J S Am Earth Sci 48:262–270
Lucchi F, Romangnoli C, Tranne A (2017) Volcanic landforms and landscapes of the Aeolian islands (Southern Tyrrhenian Sea, Sicily): implications for hazard evaluation. In: Soldati M, Marchetti M (eds) Landscapes and landforms of Italy. Springer, Cham, pp 443–453
Machin AH, Torrado FP (2005) The island and its territory: Volcanism in Lanzarote. In: Desir G, Gutiérrez F, Gutiérrez M (eds) Sixth international conference on geomorphology, field trip guide, 2, Zaragoza, pp 505–535
Manneela S, Kumar S (2022) Overview of the Hunga Tonga-Hunga Ha'apai volcanic eruption and tsunami. J Geol Soc India 98(3):299–304
Metrich N, Sigurdsson H, Meyer PS, Devine JD (1991) The 1783 Lakagigar eruption in Iceland: geochemistry, CO_2 and sulfur degassing. Contr Mineral Petrol 107:435–447
Middleton GJ, Kempe S, Webb JA (2023) Queensland lava caves. In: Webb J, White S, Smith GK (eds) Australian caves and karst systems. Springer, Cham, pp 235–244
Milner SC, Le Roex AP, O'Connor JM (1995) Age of Mesozoic igneous rocks in northwestern Namibia, and their relationship to continental breakup. J Geol Soc 152(1):97–104
Moore A, Cotterill (2010) Victoria Falls: Mosi-oa-Tunya—the smoke that thunders. In: Migoń P (ed) Geomorphological landscapes of the World. Springer, Cham, pp 143–153
Moore C, Wright T, Hooper A, Biggs J (2019) The 2017 Eruption of Erta'Ale Volcano, Ethiopia: insights into the shallow axial plumbing system of an incipient Mid-Ocean Ridge. Geochem Geophys Geosys 20(12):5727–5743
Moufti MR, Németh K (2014) The White Mountains of Harrat Khaybar, Kingdom of Saudi Arabia. Int J Earth Sci 103:1641–1643
Nelson E (2017) Montserrat. In: Allen CD (ed) Landscapes and landforms of the Lesser Antilles. Springer, Cham, pp 117–134
Newhall CG, Punongbayan R (eds) (1996) Fire and mud: eruptions and lahars of Mount Pinatubo, Philippines. Philippine Inst Volcanol Seismol, Quezon City, p 1126
Ollier CD (1988) Volcanoes. Blackwell, Oxford, p 228
Ollier CD, Sheth HC (2008) The high Deccan duricrusts of India and their significance for the 'laterite' issue. J Earth System Sci 117:537–551
Oppenheimer C, Yirgu G (2002) Thermal imaging of an active lava lake: Erta'Ale volcano, Ethiopia. Int J Remote Sens 23(22):4777–4782
Palmer HS (1927) Lapiés in Hawaiian basalts. Geogr Rev 17:627–631
Permenter JL, Oppenheimer C (2007) Volcanoes of the Tibesti massif (Chad, northern Africa). Bull Volcanol 69:609–626
Pierson TC, Janda RJ, Thouret JC, Borrero CA (1990) Perturbation and melting of snow and ice by the 13 November 1985 eruption of Nevado del Ruiz, Colombia, and consequent mobilization, flow and deposition of lahars. J Volcanol Geotherm Res 41(1–4):17–66
Pioli L, Erlund E, Johnson E, Cashman K, Wallace P, Rosi GHD (2008) Explosive dynamics of violent Strombolian eruptions: the eruption of Parícutin Volcano 1943–1952 (Mexico). Earth Planet Sci Lett 271(1–4):359–368
Ponomareva VV, Melekestsev IV, Dirksen OV (2006) Sector collapses and large landslides on Late Pleistocene-Holocene volcanoes in Kamchatka, Russia. J Volcano Geotherm Res 158(1–2):117–138
Proud SR, Prata T, Schmauß S (2022) The January 2022 eruption of Hunga Tonga—Hunga Ha'apai volcano reached the mesosphere. Science 378:554–557
Queiroz G, Pacheco JM, Gaspar JL, Aspinall WP, Guest JE, Ferreira T (2008) The last 5000 years of activity at Sete Cidades volcano (São Miguel Island, Azores): implications for hazard assessment. J Volcanol Geotherm Res 178(3):562–573
Queiroz G, Gaspar JL, Guest JE, Gomes A, Almeida MH (2015) Eruptive history and evolution of Sete Cidades Volcano, São Miguel Island, Azores. Geol Soc Lond Mem 44(1):87–104
Renne PR, Glen J, Milner SC, Duncan AR (1996) Age of Etendeka flood volcanism and associated intrusions in southwestern Africa. Geology 24(7):659–662
Richards A, Dietz R (1956) Eruption of Barcena Volcano, San Benedicto Island, Mexico. Proc Eighth Pacific Sci Congr 2:157–176
Riggs N, Ort M, Connor C, Alfano F, Conway M (2019) Volcanology and associated hazards of the San Francisco volcanic field. GSA Field Guides 55:127–146
Robinson CS (1956) Geology of Devils Tower national monument, Wyoming. Bull US Geol Surv 1021–1:289–302
Rodriguez-Gonzalez A et al. (2022) Lava deltas, a key landform in oceanic volcanic islands: El Hierro, Canary Islands. Geomorphology 416:108427
Roland NW (1974) Zur Entstehung der Trou-Au-Natron-Caldera (Tibesti-Gebirge, Zentral-Sahara) aus photogeologischer Sicht. Geol Rund 63:689–707
Sarikaya MA, Çiner A, Zreda M (2015) Fairy chimney erosion rates on Cappadocia ignimbrites, Turkey: insights from cosmogenic nuclides. Geomorphology 234:182–191
Sauro F, Pozzobon R, Massironi M, De Berardinis P, Santagata T, De Waele J (2020) Lava tubes on Earth, Moon and Mars: a review on their size and morphology revealed by comparative planetology. Earth Sci Rev 209:103288
Scarth A (1994) Volcanoes. UCL Press, London, p 273
Schmincke HU (2007) The Quaternary volcanic fields of the East and West Eifel (Germany). In: Ritter JR, Christensen UR (eds) Mantle plumes. Springer, Berlin, Heidelberg, pp 241–322
Scoon RN (2021) The geotraveller. Springer, Cham, p 398
Scott BJ, Potter SH (2014) Aspects of historical eruptive activity and volcanic unrest at Mt. Tongariro, New Zealand: 1846–2013. J Volcanol Geotherm Res 286:263–276
Sims KW et al. (2021) Mount Erebus. In: Smellie JL, Panter KS, Geyer A (eds) (2021) Volcanism in Antarctica: 200 million years of subduction, rifting and continental break-up. Geol Soc Lond Mem 55:695–739
Sinnyovsky D, Gorbach A, Gorbach V, Sinnyovska D (2023) Kamchatka—the cold and the heat of the Earth. Geoheritage 15(4):115
Sparks RS et al (1985) Ignimbrites of the Cerro Galan caldera, NW Argentina. J Volcanol Geotherm Res 24(3–4):205–248
Stevaux JC, Latrubesse EM (2010) Iguazu Falls: a history of differential incision. In: Migoń P (ed) Geomorphological landscapes of the world. Springer, Cham, pp 101–109
Summerfield MA (1991) Global geomorphology. Longman, London, p 537
Thinon I et al (2022) Volcanism and tectonics unveiled in the Comoros archipelago between Africa and Madagascar. CR Géosci 354(S2):7–34
Thouret JC (1999) Volcanic geomorphology—an overview. Earth Sci Rev 47(1–2):95–131
Thouret JC, Antoine S, Magill C, Ollier C (2020) Lahars and debris flows: characteristics and impacts. Earth Sci Rev 201:103003
Tilling RI, Topinka LJ, Swanson DA (1990) Eruptions of Mount St. Helens: past, present, and future. US Dept of the Interior, Geol Surv, Washington DC, p 56

Turab SA, Stüwe K, Stuart FM, Cogne N, Chew DM, Robl J (2023) A two phase escarpment evolution of the Red Sea margin of south-western Saudi Arabia. Insights from low-temperature apatite thermochronology. Earth Planet Sci Lett 603:117990

Vallance JW (2024) Lahars: origins, behavior and hazards. In: Jakob M, McDougall S, Santi P (eds) Advances in debris-flow science and practice. Geoenvironmental disaster reduction. Springer, Cham. https://doi.org/10.1007/978-3-031-48691-3_12

Van Gerven M, Pichler H (1995) Some aspects of the volcanology and geochemistry of the Tengger Caldera, Java, Indonesia: eruption of a K-rich tholeiitic series. J SE Asian Earth Sci 11(2):125–133

Van Westen CJ, Daag AS (2005) Analysing the relation between rainfall characteristics and lahar activity at Mount Pinatubo, Philippines. Earth Surf Proc Landf 30(13):1663–1674

Wadge G, Biggs J, Lloyd R, Kendall JM (2016) Historical volcanism and the state of stress in the East African Rift System. Front Earth Sci 4:86

Walker GPL (1990) Geology and volcanology of the Hawaiian Islands. Pac Sci 44:315–347

Whitehead PW (2010) The regional context of the McBride Basalt Province and the formation of the Undara Lava flows, tubes, rises and depressions. In: Proceedings 14th international symposium on vulcanospeleology, 2010, Undara, Queensland, pp 9–18

Widdowson M (1997) Tertiary palaeosurfaces of the SW Deccan, Western India: implications for passive margin uplift. Geol Soc Lond Spec Publ 120(1):221–248

Widdowson M, Gunnell Y (1999) Lateritization, geomorphology and geodynamics of a passive continental margin: the Konkan and Kanara coastal lowlands of western peninsular India. Spec Publ Int Assoc Sed 27:245–274

Wohletz KH, Sheridan MF (1983) Hydrovolcanic explosions II. Evolution of basaltic tuff rings and tuff cones. Am J Sci 283(5):385–413

Wood CA (1980) Morphometric evolution of cinder cones. J Volcanol Geotherm Res 7(3–4):387–413

Xu W, Ruch J, Jónsson S (2015) Birth of two volcanic islands in the southern Red Sea. Nature Comm 26:7104

Żaba J, Gaidzik K (2011) The Ngorongoro Crater as the biggest geotouristic attraction of the Gregory Rift (Northern Tanzania, Africa)—geological heritage. Geotourism 1:3–26

Závada P, Dědeček P, Lexa J, Keller GR (2015) Devils Tower (Wyoming, USA): a lava coulée emplaced into a maar-diatreme volcano? Geosphere 11(2):354–375

5 Sedimentary Rocks

Abstract

Sedimentary rocks include limestones, sandstones, conglomerates, siltstones, mudstones, shales and clays. They are formed by the accumulation or deposition of mineral or organic particles at Earth's surface. Sandstones are clastic sedimentary rocks composed mainly of sand-sized (0.0625–2 mm) silicate grains. They comprise about 20–25% of all sedimentary rock. Examples of sandstone landscapes are given from Jordan, Saudi Arabia, North Africa, Australia, South Africa, China, the American south west, and the Gran Sabana of South America. Also discussed are tors on various lithologies, and the landforms associated with conglomerates, with examples from Greece, Spain, Australia and Brazil. Landscapes formed in mudstones and other fine-grained sedimentary rocks, include Badlands. Finally, notably in Australia, silcrete duricrusts have an impact on landscapes.

Keywords

Sandstone · Quartzite · Conglomerate · Shale · Clay · Badland · Silcrete · Mud volcanoe

5.1 Introduction

Sedimentary rocks, as their name implies, are formed by a vast range of sedimentation processes (see Reading 1978; Leeder 1991). They result from the accumulation or deposition of mineral or organic particles at Earth's surface. Sedimentation is the collective name for processes that cause these particles to settle in place. The sedimentary rock cover of the continents is extensive (73% of the Earth's current land surface). The geological detritus originates largely from the weathering and erosion of existing rocks and this is transported to the place of deposition by water, wind, ice or mass movements. Biological detritus forms by bodies and parts (mainly shells) of dead aquatic organisms, as well as their faecal mass, suspended in water and slowly piling up on the floor of water bodies. Sedimentation may also occur as dissolved minerals precipitate from water as is the case of tufas (travertines) and some salt rocks. Sedimentary rocks include limestones and dolomites (see Chap. 6), sandstones, conglomerates, siltstones, mudstones, shales and clays.

5.2 Sandstones

5.2.1 Introduction

Sandstones are clastic sedimentary rocks composed mainly of sand-sized (0.0625–2 mm) silicate grains. They comprise about 20–25% of all sedimentary rock. Most sandstone is composed of quartz or feldspar. Greywackes are a variety of sandstone composed of poorly sorted angular grains. Quartzite is a metamorphic rock formed when quartz-rich sandstone or chert has been exposed to high temperatures and pressures. Such conditions fuse the quartz grains together forming a dense, hard, equigranular rock.

In this chapter we review the nature of sandstone landscapes in various parts of the world. The role of sandstones in geomorphology has been expertly reviewed by Mainguet (1972) and Young et al. (2009), while Härtel et al. (2007) provide case-studies of sandstone landscapes in Europe. Migoń et al. (2017) provide a review of the literature on rock cities, which occur in many sandstone regions but are not restricted to them.

The impact of sandstones depends in part upon their jointing/fracturing and their bedding. This is shown in the landscapes of the Kimberley in northern Australia (Fig. 5.1), which have developed in the c 1800 million year old King Leopold Sandstone.

A. Goudie, *Rocky Landscapes of the World with Google Earth*, https://doi.org/10.1007/978-3-031-77418-8_5

(a)

(b)

Fig. 5.1 **a** Fracture and joint patterns in the Kimberley district of Australia. 14°35′46.04″S125°25′38.62″E. **b** The King George River and falls in the Kimberley, Australia. The role of bedding is clear (Photo by A.S. Goudie)

5.2.2 Wadi Rum and Saudi Arabia

Cambro–Ordovician sandstones are exposed in North Africa and Arabia over an area averaging 1500 km in width and 6000 km in length, extending from Morocco in the west to the Persian/Arabian Gulf in the east (Avigad et al. 2005). For that reason we discuss in some detail sandstone terrains from Jordan, Saudi Arabia, and North Africa.

In southern Jordan, to the north of Aqaba and just to the east of the Arava depression, part of the Dead Sea rift, there is an area of magnificent sandstone terrain. Rum's most impressive landforms are huge sandstone inselbergs (Goudie et al. 2002; Migoń and Goudie 2014) (Fig. 5.2). These rise as much as 800 m above the playas and sand flats of the desert floor. Jebel Rum itself reaches an altitude of 1754 m and is the highest point in Jordan. In the neighbourhood of Jebel Rum the proportion covered by the floor is comparatively small and the landscape consists of narrow defiles between great sandstone masses, whereas north and east of Quweira the landscape is dominated by desert floor and the inselbergs are both smaller and more isolated.

The area's slopes show the impact of great rock falls (Fig. 5.2b, c) created by the undermining of cliff faces by salt weathering at their bases. There are also rock arches (Fig. 5.2d), and in addition the cliffs are notable for the extraordinary development of weathering forms, with holes, niches, columns, elephant-head shapes, drapes and the like. Similarly bizarre weathering forms, domes and clefts are displayed in the same Cambrian and Ordovician sandstones at the ancient Nabatean city of Petra (Osborn and Duford 1981; Waltham 1994). They seem to be the result of localised surface wash down the cliffs with associated carbonate precipitation and case-hardening, and of salt weathering of areas that are not washed (Viles and Goudie 2004).

There are three prime reasons why this remarkable landscape has developed. Firstly, there is a considerable thickness (around 900 m) of sandstones and these have the structural, mechanical, and lithological characteristics that enable them to produce tall cliffs and spectacular weathering forms. Secondly, there is significant local relief that has been caused by a lowering of the base-level and an uplift of the rift shoulders as the Dead Sea Basin developed in the Neogene. Thirdly, tectonic activity has produced major lineaments, including faults and fractures, and these have helped to produce some of the characteristic angularity and compartmentalisation of the landforms.

Broadly similar terrain and rocks occurs over extensive areas of neighbouring Saudi Arabia. The most familiar and iconic rocks in Al-Ula are the beautiful light-red fluvial Saq Cambrian and Ordovician sandstones. These sediments were deposited 520 to 485 mya (Fig. 5.3). These produce many of the same features seen in Wadi Rum, including arches, rock falls and cavernous weathering forms. The importance of joint control is clear.

5.2.3 North Africa

Large tracts of the Sahara are underlain by various sandstones. In the heart of this desert, in the Fezzan region of southwestern Libya, there is a sandstone area that is noted both for its rock art and for its magnificent scenery. This is the Tadrart Acacus (Zerboni et al. 2015) (Fig. 5.4). The area, with a mean annual rainfall of less than 10 mm, is well inside the hyperarid belt of the Sahara. It covers c. 5000 km^2 with a maximum elevation of c. 1300 m a.s.l. in the west. It consists of a monocline developed in Devonian sandstones (Eramo et al. 2020) and is deeply cut by a dendritic fossil drainage network of W-E-oriented wadis, which consists of deep canyons in their uppermost reaches.

One of the reasons why the geomorphology of the area is especially intriguing is that in spite of the very low rainfall, and in spite of the fact that the bulk of the rocks are siliceous sandstones, there is widespread development of karstic features. Deposits of calcareous tufa have been identified (Cremaschi et al. 2010), and underground solutional forms, reaching several hundreds of meters in depth, are well exposed. They consist of vertical and horizontal tubes, tunnels, weathering pits, large caves, rock arches, towers, and pillars. The ruiniform landscape at the eastern entrance of the massif clearly demonstrates the effects of solutional processes on sandstone. Most intriguingly of all, on its eastern margin, there are large numbers of round hills that are reminiscent of the cone karst of tropical limestone areas (see Sect. 6.6), and it is possible that they were moulded, at least in part, under tropical conditions in the Paleogene (Early Tertiary) as suggested for NE Niger by Busche (2010).

In Mali there is a famous sandstone area called Bandiagara (Fig. 5.5). This plateau, with a high east-facing escarpment, rises some 250 m from the Gondo Plain. It is composed of hard, gently dipping Silurian sandstone. The Bandiagara Plateau is continued north eastwards by the huge Douentza Massif, with the impact of jointing clearly visible (Aghassy 1970).

In the Libyan Desert in southwestern Egypt is the Gilf Kebir, a huge erosional remnant of sandstone attaining heights of more than 1000 m above sea level (Embabi 2018). It approximately equals Switzerland in size, and its south-eastern part has sheer cliffs (Fig. 5.6). These forbidding barriers have an approximate total frontage of 3000 km. The Gilf consists of two plateaux that are connected by a narrow bridge. They are composed of horizontal beds of Palaeozoic and Mesozoic sandstones that are

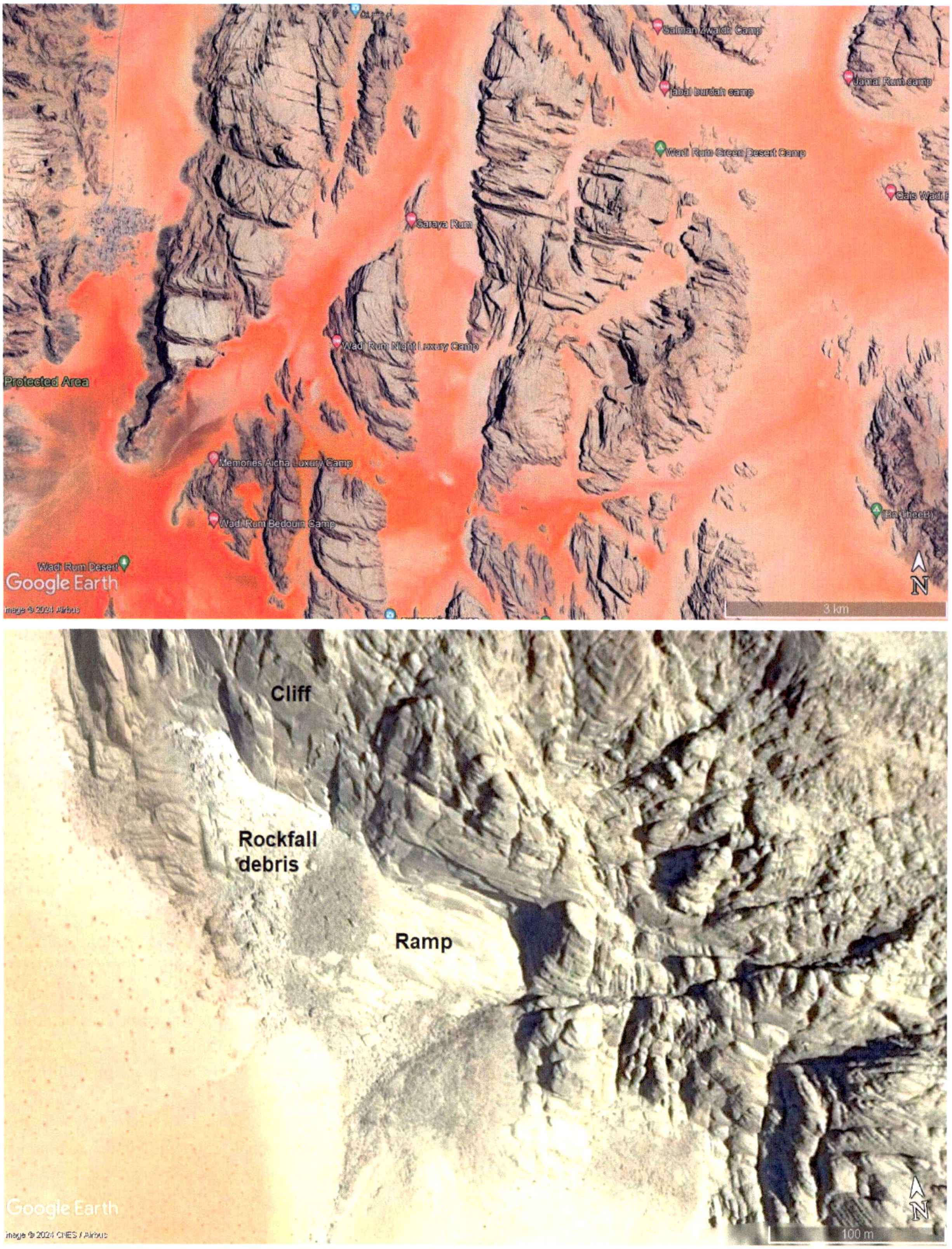

Fig. 5.2 **a** Wadi Rum, Jordan. 29°33′50.97″N35°28′17.49″E. ©Google Earth. Scale bar = 3 km. **b** Wadi Rum, Jordan, rock falls. 29°27′2.80″N 35°28′52.31″E. ©Google Earth. Scale bar = 100 m. **c** Wadi Rum rockfall (Photo by A.S. Goudie). **d** Wadi Rum arch, cliffs and defiles (Photo by A.S. Goudie)

(c)

(d)

Fig. 5.2 (continued)

Fig. 5.3 Al-Ula sandstone terrain, Saudi Arabia. 26°39′1.52″N, 37°56′51.07″E. ©Google Earth. Scale bar = 1 km

covered by a resistant caprock. The south eastern plateau is on its own about the size of Corsica. The Gilf is bitten into by a series of large, canyon-like wadis. The abrupt rise of the Gilf Kebir is made all the more remarkable in that it soars some 300 m above a land surface that is immensely flat—the 120,000 km^2 Selima Sand Sheet.

The Gilf Kebir was only discovered by Prince Kemal el Din (see Goudie 2008) in 1926 and was the subject of detailed analysis during a 1938 expedition by two British geomorphologists, Ralph Bagnold and Ron Peel. The plateau was described thus by Peel (1941, p. 8).

> This plateau... has a total length north to south of about 150 miles, a maximum breadth of some 50 miles, and a total area about equal to that of Wales. On all sides except the north it is sharply separated from the surrounding plains by abrupt cliffs, having average slopes of about 35°. These cliffs are highest in the south and west, where they reach above 1 100 feet. To the north the height decreases, and the northernmost extensions of the plateau merge into the general plains. Long branching wadis have cut deep into the heart of the plateau, and in one or two places have cut right across it. In some places, particularly to the west, the desert plains extend unbroken up to the base of the cliffs, but at others the plateau is screened behind a maze of detached fragments and dissected hills.

To Peel, the outstanding feature of the Gilf was the abrupt distinction between desert floor and plateau surface: the two levels seemed to be entirely distinct, with no transition between them. He believed that erosional sapping by groundwater flow had played a major role in the formation of the Gilf, an area that has frequently been used as an analogue for Mars (Maxwell 1980). Numerous outliers show that the Plateau was formerly more extensive (McCauley et al. 1982a). Large fossil drainage systems, fed by formerly higher discharges, run from the Gilf into the Great Sand Sea in the north and into north-west Sudan to the south (McCauley et al. 1982b).

A very large sandstone massif is Ennedi in Chad (Fig. 5.7). This, a World Heritage Site, has an area of approximately 60,000 km^2 (as large as Switzerland), and its highest point is approximately 1450 m above sea level. It is composed of Cambrian-Ordovician to Carboniferous sandstones overlying a Precambrian granite base. Major joint systems create a maze of deep defiles and isolated blocks of cliffed terrain, arches, pillars, mushroom rocks, sand-filled alluvial wadi channels, and in exceptional cases, permanent water-filled pools

(a)

(b)

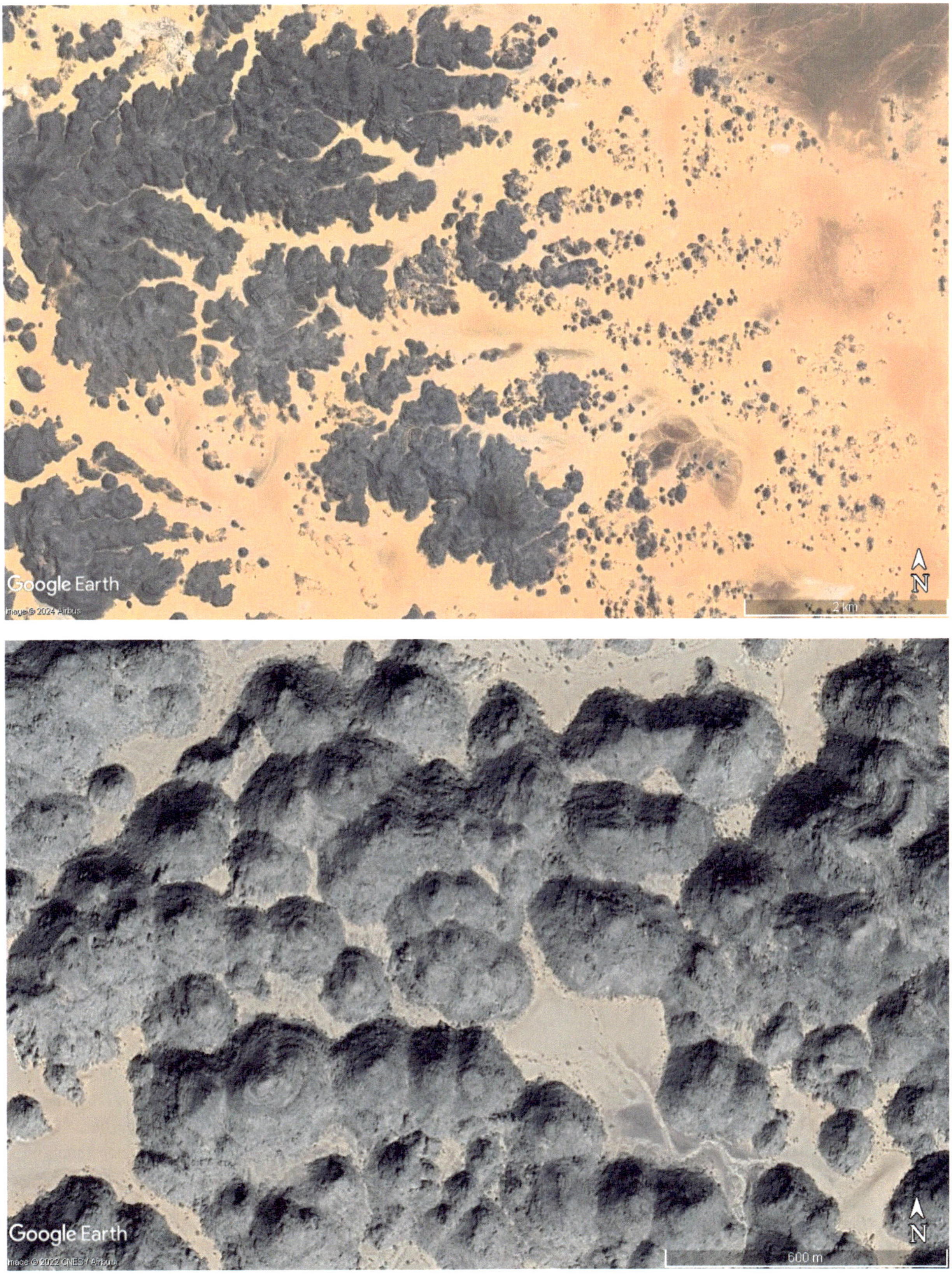

Fig. 5.4 **a** Acacus, Libya. 24°58′15.25″N, 10°42′55.33″E. ©Google Earth. Scale bar = 2 km. **b** Eastern margins of the Acacus massif, Libya, showing conical hills and closed basins which may be Tertiary (Paleogene?) karst relics. 25°14′N, 10°34′E. ©Google Earth. Scale bar = 600 m. **c** Sandstone weathering phenomena in the Tadrat Acacus. (Photo by H.A. Viles)

(c)

Fig. 5.4 (continued)

(*gueltas*) fed by springs (Knight 2024). Le Heron (2018) has identified one unusual feature of Ennedi on satellite images, namely the presence of an array of curvilinear structures. These were interpreted as exhumed mega-scale glacial lineations associated with Palaeozoic paleo–ice stream pathways.

Some desert sandstone terrains have a suite of remarkable and, possibly, unexpected landforms. This is the case of the Djado Plateau in the Sahara Desert in north east Niger, which in spite of the areas's profound aridity (mean annual rainfall c 20 mm), has a series of colossal landslides and a suite of karstic forms (Fig. 5.8). It is a maze of low escarpments, sand-choked canyons and wind-carved outcrops, which is geologically contiguous with Algeria's Tassili N'Ajjer and Libya's Akakus. The bedrock is Nubian Sandstone. Djado has been the subject of intensive research by Detlef Busche (Busche 2010).

The fossil landslides consist of deep-seated and rotational forms (Schoeneich and Bouzou 1996). These are well displayed on the scarp and valley sides of Mangueni (Fig. 5.9). It is probable that these, together with associated mudflows, were formed during moister phases of the Early Pleistocene. Busche (2001) argued that they developed after an initial arid phase had caused oversteepening of stable slopes formed under a humid climate during the Neogene. He thought that they post-date the dissection of Late Tertiary pediments and are older than all fluvial terraces, lake sediments and aeolian corrasion and deposition features of the central Sahara. Even earlier, silcretes and ferricrete duricrusts were formed, and deep-reaching karstification took place. More recently, some of the Palaeozoic sandstone expanses of the Sahara have been moulded into mega-yardang fields by aeolian erosion, as is the case with Borkou in Chad (Goudie 2007) (Fig. 5.10).

Fig. 5.5 Douentza Massif, Mali. 15°9′36.30″N, 2°38′29.35″W. ©Google Earth. Scale bar = 20 km

In quite extensive parts of Africa, western India (see Sect. 4.7) and Australia, there are spreads of ancient laterites (ferricretes), which are a type of iron-rich duricrust. One example comes from Kordofan in Sudan, where Miocene laterite, which originally developed under humid conditions and formed in a low-lying area, has now, as a result of relief inversion, created a plateau capping near Jebel Howag (Schwarz 1994) (Fig. 5.11). Laterite-capped plateaux are common in northern Nigeria (Yakubu and Ojanuga 2013) and French workers refer to the landscape impact of *cuirasses* in West Africa (Tricart 1972, Chap. 4). Laterites have high infiltration capacities and shear strengths, which make them resistant to erosion, and their induration of different landform elements (e.g. river terraces, pediments, and erosion surfaces) can in parts of West Africa lead to landscapes that take on 'the appearance of giant staircases' (Michel 1969, p. 126). In Guinea, bauxite crusts can play a similar role (Chardon et al. 2006). The presence of laterite profiles in which there are marked differences between the surface capping and the more friable and finer material beneath can favour the formation of pseudo-karstic phenomena produced by sub-surface flushing. Cave formation and roof collapse are therefore not uncommon (Bowden 1980), and soup-plate like forms can be produced.

5.2.4 Australia

The Bungle Bungle Range is a major landform and the main feature of the Purnululu National Park located in the Kimberley region of Western Australia. It is a World Heritage Site and is composed of Devonian-age Elder Sandstone that has been dissected into a series of cones or domes with convex to ovoid summits. These are believed to be superlative examples of beehive-shaped sandstone karst rising 250 m above the surrounding semi-arid savannah grasslands. Their spectacular black and orange banded appearance has been formed by biological processes of cyanobacteria (single cell photosynthetic organisms) which serve to stabilise and protect the ancient sandstone formations. The intricate maze of towers is accentuated by

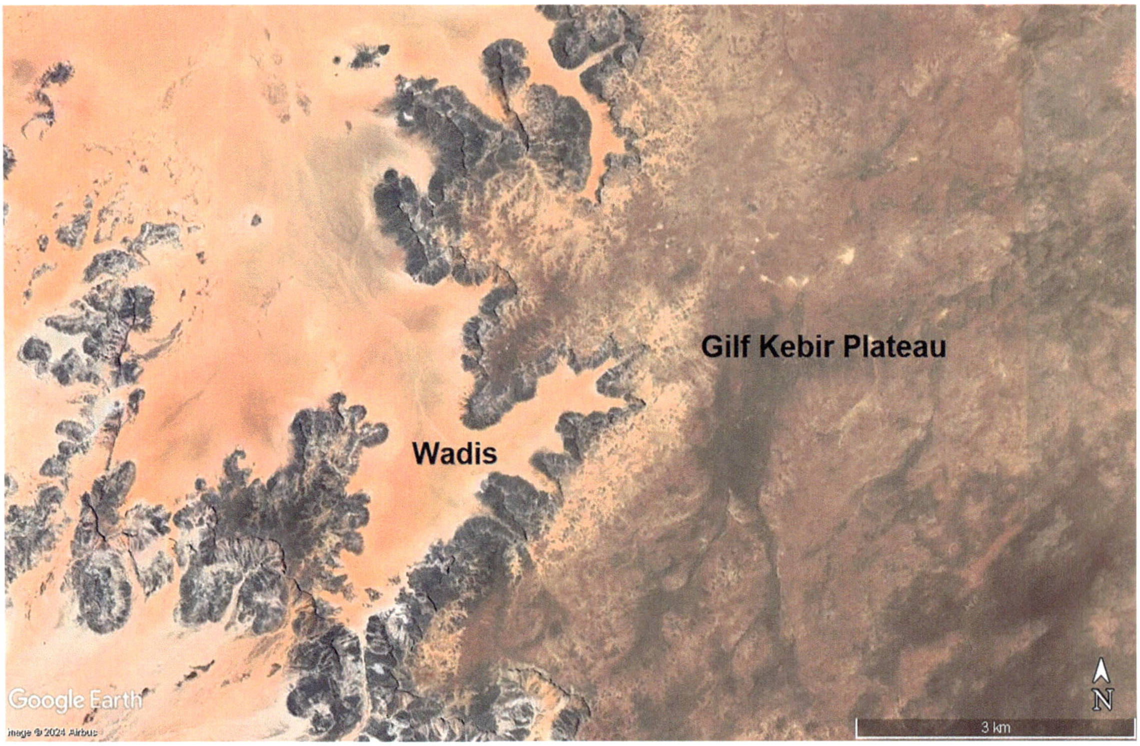

Fig. 5.6 Gilf Kebir, Egypt. 23°22′39.83″N, 25°47′47.35″E. ©Google Earth. Scale bar = 3 km

Fig. 5.7 Heavily-jointed sandstone, Ennedi, Chad. 16°43′39.98″N, 22°0′29.81″E. ©Google Earth. Scale bar = 1 km

Fig. 5.8 Djado Sandstone, Niger. 20°56′32.87″N, 12°23′0.55″E. ©Google Earth. Scale bar = 100 m

sinuous, narrow, sheer-sided gorges and box valleys lined with fan palms (Fig. 5.12). The massif has been exposed to weathering for at least 20 million years, and during that time may have been subject to humid tropical weathering (Young 1986).

Young (2010) attributed the spectacular tower and ridge assemblages of the Bungle Bungles to the fact that the sandstones have unusual mechanical properties which enable the sandstone to stand in steep faces but to be easily eroded. This is due to the almost complete absence of natural cement between the quartz grains. The sandstones have high compressive strength but low shearing and tensile strengths.

Uluru (Ayers Rock), in the Northern Territory of Australia, is perhaps the best known single landform in Australia and is among the best known in the world (Twidale 2010; Bourman et al. 2015) (Fig. 5.13). This steep-sided dome is a remarkable feature of great visual impact and with intriguing sculptural details, but it is also of remarkable antiquity.

Its summit stands rises some 350 m above the desert plains. Unlike many inselbergs, it is not made of granite but of almost vertically dipping (80–85°) Cambrian arkosic sandstone (the Mutitjulu Arkose). This is made evident by the parallel grooving seen in the image. The Mutitjulu Arkose is part of a c 2.5 km thick succession of sediments that were originally near horizontal and buried by 100s–1000s of metres of younger sediments. It has been tilted at c 80° to the southwest in such a way that the younger part of the sequence lies on the southwestern end of Uluru, with the oldest part of the exposed Mutitjulu Arkose occurring near Taputji (Little Ayers Rock). A network of joints subdivides Uluru, but the joints are few, widely spaced and so tightly closed that erosion is inhibited. This means that the rock has structural integrity and so the inselberg behaves essentially as a single rock mass (Bourman et al. 2015). The inselberg cannot simply be explained in terms of either lithology or faulting, for two reasons: first, similar bedrock underlies the surrounding plains and, secondly the inselberg's steep sides are not fault scarps.

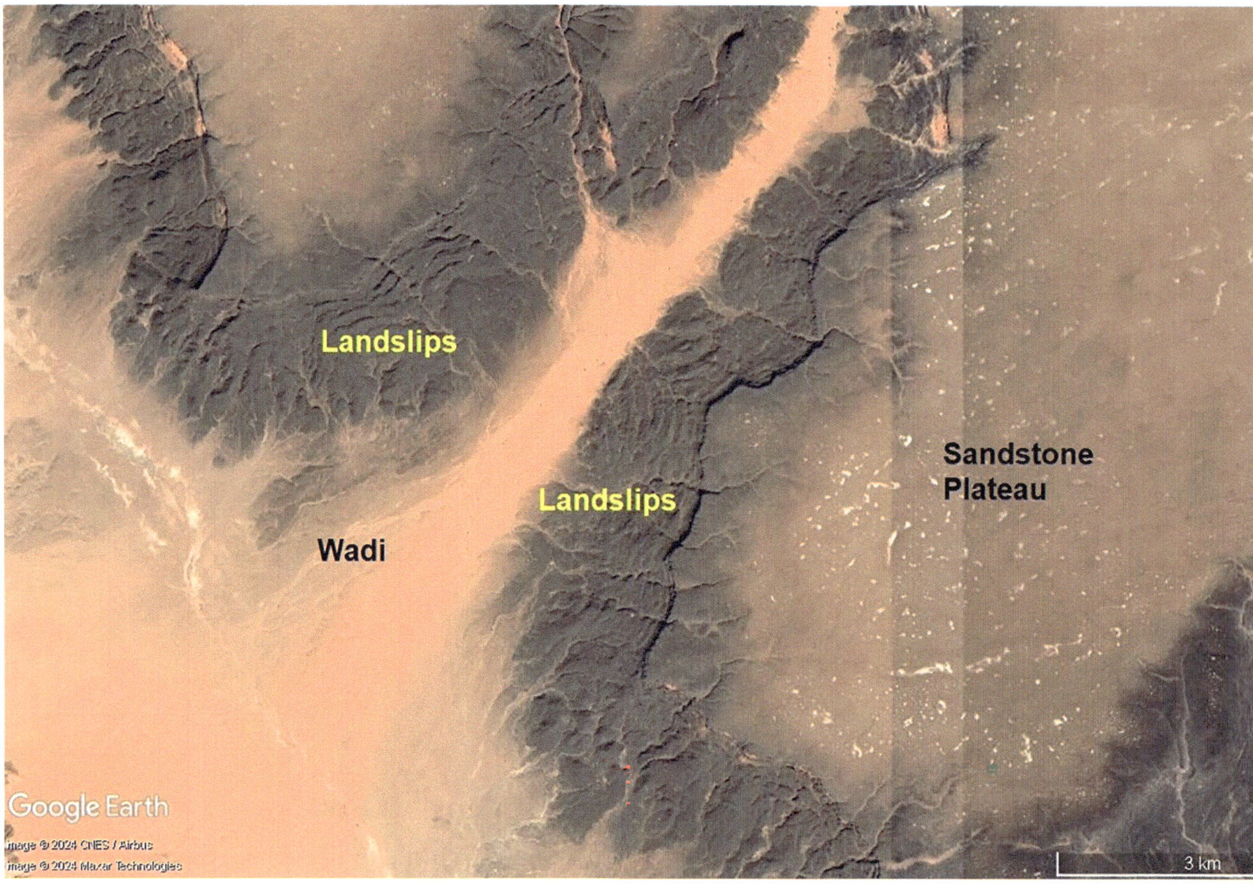

Fig. 5.9 Mangueni, Niger. 22°21′56.26″N, 12°37′23.30″E. ©Google Earth. Scale bar = 3 km

Uluru is a residual of a great thickness of rock and it is located in a synclinal structure. It may therefore have been strengthened by compressive forces either during middle Palaeozoic orogenesis or as a consequence of later warping. Its age is uncertain (Twidale 2010), but it may have evolved as a relief feature as far back as the Cretaceous. Its flanks have been steepened by scarp foot weathering, undermining and collapse. Sheet structures occur, including the famous Kangaroo's Tail. When rain storms occur, streams run off the dome, creating waterfalls and potholes on its flanks. The spalling that has taken place on its slopes has sometimes been attributed to chemical weathering such as hydration. It is possible, however, that some of the oversteepened lower slopes on the lower portion of the inselberg may be the result of spalling caused by fire (Buckman et al. 2021).

5.2.5 Golden Gate, South Africa

Sandstones are widespread in South Africa, with the landscapes associated with the Table Mountain Sandstone being particularly famous. In the Free State lies the Golden Gate Highlands National Park. The prime attraction of this park is the weathering phenomena of the fine bluffs and cliffs formed in the Clarens Sandstone (Grab et al. 2011) (Fig. 5.14). This rock, formed when the area was a desert 190–196 mya, is composed nearly entirely of fine to medium-grained, thickly-bedded sandstones. Case hardening and surface spalling occur, and some sandstone surfaces have been modified by lichen attack. There are some arches, pedestal rocks, cavernous weathering forms, rock basins, rock doughnuts, and polygonal cracking.

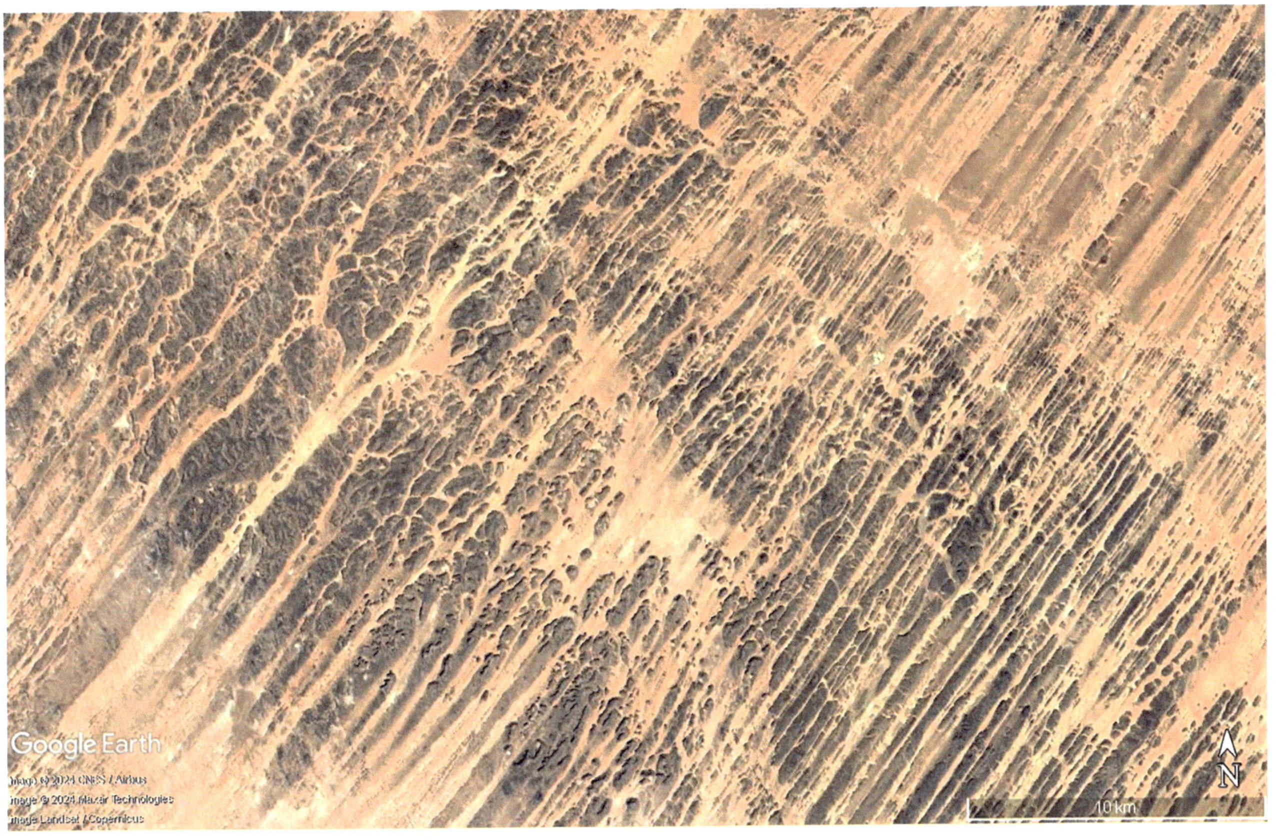

Fig. 5.10 Borkou sandstone yardangs, Chad. 18°54′43.87″N, 19°9′52.17″E. ©Google Earth. Scale bar = 10 km

5.2.6 China

The Bingling area, China, noted for its Buddhist grottoes, is located in the Dasi Valley watershed, a sub-catchment of the Yellow River, in Gansu Province. The geological strata to which the grottoes are attached belongs to the Lower Cretaceous Hekou Group, which is up to 300 m thick. The lithologies in the area consist of interbedded red sandstone (medium-fine-grained arkose) and conglomerate (mostly fine-grained). The area is notable for its steep cliffs, towers and cavernous weathering forms (Liao et al. 2024) (Fig. 5.15).

Zhangjiajie, a UNESCO Global Geopark, is located in Hunan Province in the middle of the Yangtze River basin in China (Brierley et al. 2011). It has a humid, monsoon climate. The Zhangjiajie Sandstone Peak Forest Geopark covers a total area of approximately 398 km^2 and is situated near the junction of the southwest Yunnan-Guizhou Plateau and the Wuling Mountains. The thickness, age and uniformity of its Upper and Middle Devonian sandstone beds, along with the density, angularity and depth of its joint pattern, have been crucial determinants of the existence, morphology and stability of more than 3100 thin and very high vertical sandstone pillars and peaks. This landscape type is called 'quartz sandstone peak forest' (Yang et al. 2012) (Fig. 5.16). The height, narrowness, cragginess, and number/density of sandstone columns, pillars and peaks is remarkable. Some blocks extend up to 350 m high, while more than 1000 are higher than 200 m. The peaks show a remarkable average density of 37.5 peaks per/km^2. Because of their strong cementation with iron and siliceous cement, these sandstones are very stable and strongly resistant to chemical weathering. They have a high compressive strength and commonly constitute the solid bases of the cliff walls and peak columns.

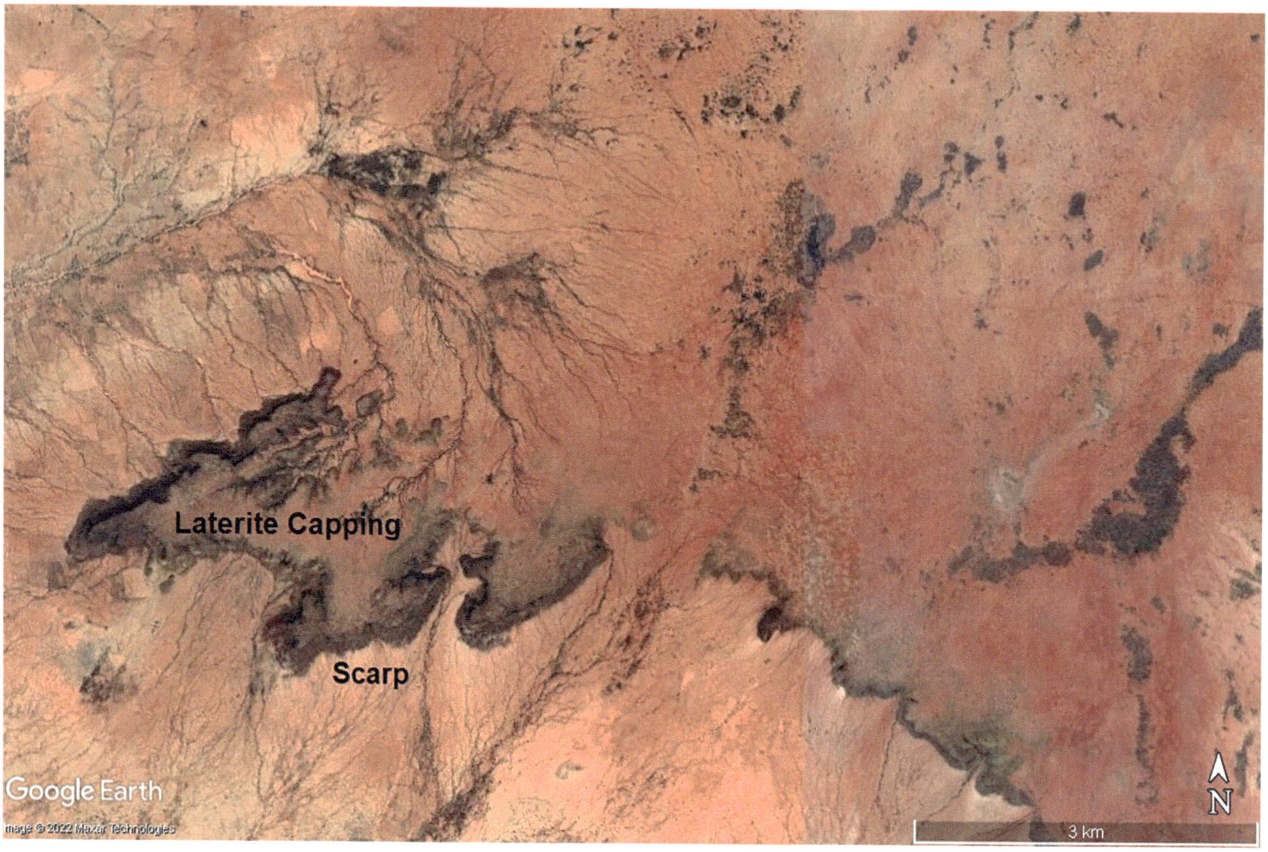

Fig. 5.11 Kordofan laterite-capped hills, Sudan. 12°31′10.90″N, 28°45′22.62″E. ©Google Earth. Scale bar = 3 km

5.2.7 The American West

Natural arches are formed when weathering, together with mass collapse, and in arid areas with wind erosion, creates a tunnel through a slab of rock. They can thus be distinguished from natural bridges which are formed by fluvial or marine erosion. They most commonly occur in sandstone, which has sufficient permeability to provide the seepage that promotes weathering, yet which has the necessary cohesion for an arch to develop. Arches are most numerous, as in the Arches National Park in Utah, where long and closely spaced joints have been eroded to form narrow fins of rock that are readily pierced by weathering (Dixon 2010). These characteristics, and thus a very high concentration of arches, occur in the Entrada and Cedar Mesa Sandstones of the Colorado Plateau. While they can occur in granites (as in the central Namib) and in chalks (as in the Western Desert of Egypt), they are best developed in sandstones, as in Wadi Rum in Jordan. Of the ten largest natural arches in the world, all are in sandstones.

The area around Arches and Canyonlands National Parks in Utah has 'the highest concentration of natural arches in the United States and perhaps the world' (Blair 1987, p. 597). Dixon (2010) reports that the area has in excess of 700 arches ranging in height from 1 to 93 m in an area of c 300 km^2. Underlying the Jurassic Entrada Sandstone is an anticline containing salt, and the collapse of this caused the joints and fractures to form (Fig. 5.17). The arches are developed where the sandstone strata are horizontal or gently dipping. The arches were formed by a unique set of circumstances involving Middle Pennsylvanian (about 308 Ma) to Late Triassic (200 Ma) movement of sub-surface salt layers, Middle Pennsylvanian to Late Cretaceous (about 70 Ma) deposition, and Tertiary and Quaternary (23 Ma to the present) folding, faulting, erosion, and salt dissolution (Thomas and Willis 2019). The Slick Rock Member of the Entrada Sandstone, which is normally 60–110 m thick, is perhaps the most important lithological unit, for the majority of the arches occur along its lower and upper contacts and along the easily weathered beds

(a)

(b)

Fig. 5.12 **a** Bungle Bungles, Western Australia. 17°29′24.56″S, 128°22′22.90″E. ©Google Earth. Scale bar = 300 m. **b** Bungle Bungles from a light aircraft (Photo by A.S. Goudie)

Fig. 5.13 Ayers Rock (Uluru), Australia. 25°20′41.52″S, 131°2′14.52″E. ©Google Earth. Scale bar = 1 km

in the middle of the unit. All of Utah's largest arches have developed in massive quartz sandstone. Its homogenous nature allows for large conchoidal (curved) fractures that are required to produce the inherently stable arch shape. Full details of the locations and geometry of fins in the Colorado Plateau are given in Barnes (1987), and Munthe (2002) catalogues the arches of the Escalante Canyons and Kaiparowits Plateau of southern Utah.

The Colorado Plateau, which covers a vast area of Utah together with parts of Arizona, Colorado and New Mexico, is an uplifted geologic terrain characterized my mostly flat-lying Upper Palaeozoic and Mesozoic sedimentary rocks with significant lithological differences that control the nature of its geomorphology. It has some very dramatic sandstone landscapes, and one landform type that is widespread is the alcove. These are found, for example, in Zion Canyon. They were also favoured sites of Indian pueblo settlements, such as Mesa Verde (Fig. 5.18). They appear to form in massive sandstones, such as the Jurassic Navajo Sandstone, at least in part as a result of groundwater sapping, pressure release, and associated salt weathering (Laity and Malin 1985; Howard et al. 1988). Especially where joints are widely spaced, the hollow developed in a cliff may not penetrate through the rock mass. Instead of a true arch, an alcove or apse develops. These are much more widespread than arches, but they form in essentially the same manner.

Zion Canyon was established as a National Park in 1919. Its canyons, cliffs and alcoves were sculpted from the Navajo Sandstone, a rock that was formed in a desert environment in the Jurassic (Fig. 5.19). Subsequent uplift of the area led to drainage incision by the Virgin River and its tributaries and to fracturing of the rock. Many of the slopes in Zion Canyon display exfoliation features, a common phenomenon in the sandstones of the Colorado Plateau. This large-scale exfoliation is similar to that found in granite landscapes and it has been produced in the same way—by differential residual stresses resulting from erosion, particularly canyon cutting, of rocks that were once buried (Bradley 1963) (see Sect. 3.5).

Nearby Bryce Canyon (Fig. 5.20) has very different scenery and rock types from Zion Canyon. It is known for

Fig. 5.14 **a** Golden Gate National Park, South Africa. 28°29′52.88″S, 28°35′16.74″E. ©Google Earth. Scale bar = 500 m. **b** Clarens Formation cliffs at Golden Gate, South Africa (Photo by A.S. Goudie)

(a)

(b)

Fig. 5.15 **a** Bingling, China. 35°47′51.28″N 103°3′40.69″E. ©Google Earth. Scale bar = 900 m. **b** Bingling, China (Photo by A.S. Goudie)

(a)

(b)

Fig. 5.16 **a** Zhangjiajie, China. 29°19′45.83″N, 110°25′56.70″E. ©Google Earth. Scale bar = 400 m. **b** Zhangjiajie peak forest (Photo by A.S. Goudie)

Fig. 5.17 **a** Moab sandstone fins. 38°32′17.12″N, 109°34′30.43″W. ©Google Earth. Scale bar = 1 km. The Colorado River is on the left of the image. **b** Landscape Arch. Located in the Devils Garden area of Arches National Park, it has an incredible span of 88 m and is remarkable due to its gravity-defying ribbon of rock that in places narrows to only 2 m in thickness. The arch is entirely within a fin of the Slick Rock Member. It may not endure much longer—two large pieces fell off in 1991 and 1995. An alcove (proto-arch) occurs in the middle distance (Photo by A.S. Goudie)

(a)

(b)

Fig. 5.18 **a** Mesa Verde, USA. 37°10′55.24″N, 108°29′19.60″W. ©Google Earth. Scale bar = 700 m. **b** Pueblo Village, Mesa Verde, USA (Photo by A.S. Goudie)

(a)

(b)

Fig. 5.19 **a** Zion Canyon, Utah, USA. 37°12′17.88″N, 112°57′8.09″W. ©Google Earth. Scale bar = 3 km. Note the fractures in the Navajo Sandstone. **b** Zion Canyon. A steep cliff of resistant Navajo Sandstone with alcove development and exfoliation features (Photo by A.S. Goudie)

(a)

(b)

Fig. 5.20 **a** Bryce Canyon, Utah, USA. 37°36′52.67″N, 112°8′56.15″W. ©Google Earth. Scale bar = 1 km. **b** Bryce Canyon, Utah, USA (Photo by A.S. Goudie)

distinctive hoodoos, spires and towers that appear as forests of rock. The hoodoos are 'rock columns, chimneys or pinnacles possessing bulging knobs and narrow recesses' (Morris et al. 2012, p. 62). It was established as a National Park in 1928 to help preserve the intricately eroded and brightly coloured amphitheatres that beset its rim. The Eocene Claron Formation, which includes limestone strata as well as sandstone and shale, is the main rock formation in the amphitheatres. These rocks are relatively soft and erodible, and are subject to weathering by solution (in the case of the limestone members) and freeze–thaw action. In the winter months, melting snow seeps into cracks and joints and freezes at night. Over 200 of these freeze/thaw cycles occur each year. Erosion occurs in response to summer rainstorms. Internal layers of mudstone, conglomerate and siltstone interrupt the limestone horizontally. These layers are more resistant to solutional and other forms of attack and can therefore act as protective capstones of fins, windows and hoodoos. Many of the more durable hoodoos are capped with dolomite (Morris et al. 2012). In short it is the weathering and erosion of a stack of relatively weak rocks of different types that creates the area's distinctive morphology.

It is remarkable that Monument Valley in the Navajo National Park, one of the most famous of all sandstone landscapes in the American west, has been the subject of rather little detailed geomorphological research (Fig. 5.21). Its magnificent scenery is in part related to the areas's presence astride a great upwarped area—the Monument Upwarp—that occurs across the Utah-Arizona border. This has caused rivers to adjust by cutting down and undertaking widespread erosion that has led to the rock remnants that compose Monument Valley. The second main influence is the presence of a sequence of sedimentary rocks of different lithologies. The prominent cliff-forming formation is the De Chelly Sandstone, which is of Middle Permian age. This is being gradually eroded back from the underlying, more erodible, Organ Rock Shale. The backward erosion has created the mesas, buttes, and pinnacles, which are also capped by the Triassic Moenkopi Sandstone and by a thin remnant of the Shinarump Conglomerate of the Chinle Formation (Baars 2000).

Some Entrada Sandstone areas display remarkable weathering pits. Two key sites are at Cookie Jar Butte and Rock Creek Bay in Utah (Netoff and Shroba 1993). Depths of up to 16.5 m are found and at Rock Creek Bay the long axes of some coalesced pits commonly exceed 35 m, with the longest one measuring 72 m (Fig. 5.22). It needs to be stressed, however, that these are exceptionally large examples, and most of those reported in the literature are very much smaller (Fig. 5.22c). The processes involved in pit formation include solution, hydration, biochemical weathering, and deflation.

5.2.8 The Gran Sabana Quartzite Landscapes

Quartzites are metamorphosed sandstones. They can be very resistant to erosion and so may be associated with steep escarpments, as is the case in the Blyde River canyon in eastern South Africa (24°34′, 30°49′E), and the Serra do Espinhaço in Brazil (17°41′S, 43°45′W) (Magalhães Junior et al. 2015). Cosmogenic dating indicates that over long time spans they undergo low rates of fluvial denudation (Peifer et al. 2021). Paradoxically, however, they may also display some striking pseudo-karstic forms.

In the Gran Sabana on the borders of Venezuela, Guyana and Brazil is a spectacular mesa and butte topography with well-developed karst forms (Wray 2010; Doerr 1999) (Fig. 5.23), including those of the Canaima National Park in Venezuela (https://earthobservatory.nasa.gov/images/148520/venezuelas-flat-topped-mountains) (Accessed 25th June 2024). These are hosted by Pre-Cambrian Roraima Group quartzites and sandstones. The local name for the mesas or table-mountains is *tepui*. Some of the mesas tower over 1000 m above the lowlands and range in size from a fraction of a km^2 to as much as 15 km^2. Cascading off one of the tepuis in Canaima are the Angel Falls, the world's tallest uninterrupted waterfall, with a height of 979 m and a plunge of 807 m. The quartzite karst landforms, produced under high rainfall conditions, are comparable to those developed in carbonate rocks and include such features as karren, sinkholes, poljes, ruiniform relief, and caves. The Sarisariñama massif in Venezuela hosts the world biggest collapse dolines in quartz-rich lithologies, with volumes up to some millions of cubic meters (Sauro et al. 2019). Yanes and Briceño (1993) attribute this to intense chemical weathering under humid tropical conditions, together with the constant washing out and removal of fine-grained detritus.

5.3 Quartzite, Sandstone and Gritstone Tors

In some parts of the world, quartzites produce upstanding landforms in the shape of tors. In Shropshire, England, there is an area called the Stiperstones. Here Ordovician

(a)

(b)

Fig. 5.21 **a** The mesas and buttes of Monument Valley, USA. 38°16′18.13″N, 110°55′22.33″W. ©Google Earth. Scale bar = 3 km. **b** Monument Valley, USA (Photo by A.S. Goudie). This image shows the West Mitten butte (left) with its distinctive pinnacle

(a)

(b)

Fig. 5.22 **a** Giant sandstone weathering pits in Entrada Sandstone at Rock Creek Bay, Utah. 37°10′12.60″N, 111°11′53.53″W. ©Google Earth. Scale bar = 80 m. **b** Giant sandstone weathering pits at Cookie Jar Butte, Utah. ©Google Earth. 37°5′14.83″N, 111°17′45.75″W. Scale bar = 100 m. **c** Ground shot of smaller weathering pits in Utah (Photo A.S. Goudie)

(c)

Fig. 5.22 (continued)

quartzites rise up above valleys that are mainly developed in less resistant shales (Goudie and Piggott 1981; Goudie and Gardner 1992, p. 68) (Fig. 5.24). The slopes leading away from the tors are covered in angular debris. This occurs as a series of stripes and polygons and such patterned ground occurs in cold, often tundra environments. At the time of the last Pleistocene glaciation the area was on the margins of a great ice cap, and this produced conditions that led to frost shattering of the tors and the movement and sorting of this debris under periglacial conditions.

Tors are common features elsewhere in England. The granite ones of the south west Peninsula are particularly well known, but there are some areas where they have developed in sandstones and gritstones as at the Roaches (Fig. 5.25). These lie in the Staffordshire Peak District close to Leek. The Roaches, together with neighbouring Hen Cloud and Ramshaw Rocks, are formed from a thick bed of coarse sandstone (gritstone) of Carboniferous age, which formed from ca 315 to 326.4 mya.

Other gritstone features, including tors and ruiniform relief occur at the Brimham Rocks in Yorkshire (Duszyński and Migoń 2020a). No less striking are the landforms developed in the Lower Cretaceous Ardingly Sandstones of the High Weald in south east England (Duszyński and Migoń 2020b).

5.4 Conglomerates

5.4.1 Introduction

Conglomerates are a type of sedimentary rock that is composed at least 30% of rounded to sub-angular clasts larger than 2 mm in diameter (e.g. boulders, cobbles, pebbles, granules). They can be found in sedimentary rock sequences of all ages but probably make up less than 1 percent by weight of all sedimentary rocks. Typically they contain a matrix of finer-grained sediments, such as sand, silt or

(a)

(b)

Fig. 5.23 **a** Tepuis (mesas) at Roraima on the borders of Venezuela, Guyana and Brazil. 5°12′49.95″N, 60°45′49.31″W. ©Google Earth. Scale bar = 10 km. **b** Tepuis in the Canaima region of Venezuela. 5°16′12.11″N, 62°7′34.32″W. ©Google Earth. Scale bar = 10 km. Note the importance of joint control

Fig. 5.24 Quartzite tors and periglacial stripes Stiperstones, Shropshire, England. 52°34′52.26″N, 2°56′6.10″W. ©Google Earth. Scale bar = 100 m

clay. This fills the interstices between the clasts. The clasts and matrix are typically cemented and lithified by such substances as calcium carbonate, iron oxide or silica.

Conglomerates can create some striking and distinctive landforms and these have been expertly reviewed by Migoń (2020a, b). Among these are steep cliffs, domes with convex tops, deep and narrow linear incisions, tors, and in the case of limestone conglomerates, some karstic features. Migoń (2020a, b, Sect. 3.3) points out that 'although cliffs are so pervasive in some conglomerate terrains, top parts of bare elevations built of well-lithified conglomerates are often strikingly rounded. They assume the shape of a dome, with very steep slopes in the lower and middle sections, gradually declining in angle while going upwards, until a convex hilltop is attained'. Tors also occur. One such area noted by Migoń is at Peracense Castle in Aragon, Spain, where tens of conglomerate tors are present, rising from the gentle slopes or the steep front of an escarpment. The conglomerates are of Triassic age (Gutierrez et al. 2005) (Fig. 5.26).

Migoń (2020a, b, Sect. 3.4) also remarks, 'Deep and narrow linear incisions are common in many conglomerate terrains. Although they are quite clearly related to structural discontinuities within the rock mass (fractures, faults), they are not necessarily all of the same origin. Both weathering and fluvial processes may play a role, as well as mass movements leading to joint opening.' A good example, is Serra da Capivara in Brazil (da Silva Mutzenberg et al. 2015).

Some landforms can be composed both of sandstones and conglomerates. This is the case with the much-studied Danxia scenery of China (Yan et al. 2015), mainly distributed in south eastern, south western and north western parts of that country. Literally in Chinese, Danxia means 'reddish rays' or 'rosy cloud' during a sunrise or sunset. Danxia landforms are defined as a landscape type that has developed in red terrestrial clastic rocks, especially sandstones and conglomerates (Peng 2020) in locations like Longhushan. The distinctive features of these striking landforms are bare red walls and cliffs, with stone pillars, but many examples appear to be heavily gullied areas incised into red strata (Fig. 5.27). The Zhangye area shown in the image is a UNESCO Global Geopark. In China, the exposure of red beds covers 9.16 × 105 km^2, which accounts for

(a)

(b)

Fig. 5.25 **a** Gritstone tors, Roaches, Staffordshire, England. 53°10′25.02″N, 2°0′4.93″W. ©Google Earth. Scale bar = 30 m. **b** The Roaches, Staffordshire, England

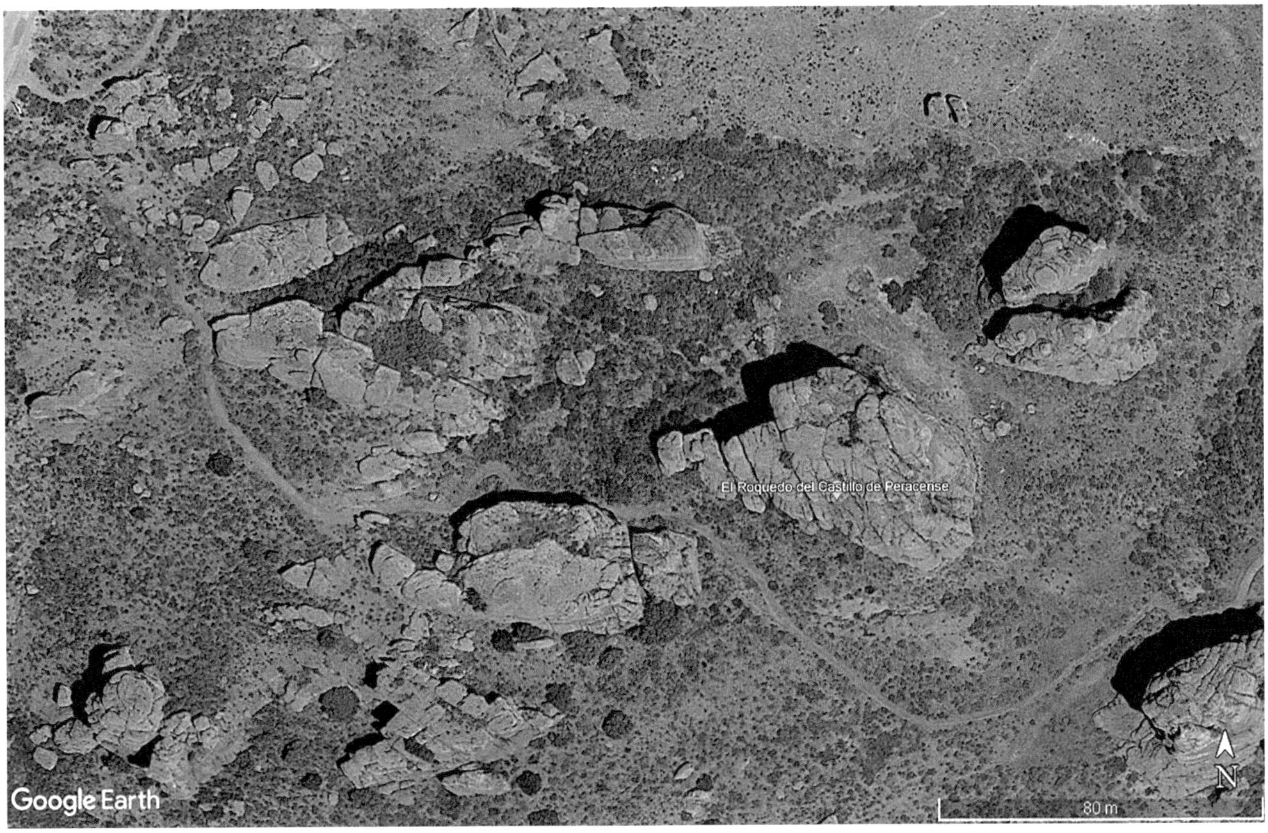

Fig. 5.26 Conglomerate tors at Peracense, Spain. 40°38′7.30″N, 1°28′46.11″W. ©Google Earth. Scale bar = 80 m

9.5% of the total land area of the country. Red beds were deposited from the Triassic to the Neogene. Among them, the Cretaceous accounts for 57% of the total area of red beds, Jurassic for 25%, Neogene for 4.8%, Paleogene for 5.1%, and Triassic for 4% (Yan et al. 2019).

5.4.2 Examples

In Meteora, Greece, on the eastern flank of the Pindus Mountains, there are 24 giant pinnacles that average about 300 m in height and which are formed of conglomerate. It is a UNESCO World Heritage Site (Fig. 5.28). The conglomerate consists of wedge-shaped masses of deltaic material with large cross-set bedding, into which were cut deep channels containing giant bars of cobbles and sand. This conglomerate, Oligo-Miocene in age (Ori and Roveri 1987; Rassios et al. 2020), has been incised into towers and buttes and much of this dissection has been located along steeply dipping joints (Young et al. 2009, p. 36). Meteora is famous for the monasteries that occur on some of its peaks.

Also in Mediterranean Europe, there are other conglomerate pinnacles. An example of a conglomerate landform can be seen at Montserrat near Barcelona in Spain (Fig. 5.29). This isolated massif is formed by thick layers of conglomerate interleaved by siltstone/sandstone from a Late Eocene fan-delta. Here, erosion has created vertical channels that give the characteristic jagged shapes the mountain is named for - Montserrat literally means 'jagged mountain'. The relief is around 100–150 m, and the influence of regional jointing pattern is evident. Dominant joints strike SSW–NNE. The steep slopes of the mountains are very prone to landslips (Janeras et al. 2017).

Another imposing conglomerate mass occurs at Los Mallos de Riglos (Fig. 5.30). These are a set of Miocene conglomerate rock forms, located in Aragon, also in Spain (Guerrero et al. 2014).

In Australia, some 30 km to the west of Uluru (Ayers Rock), is Kata Tjuta (the Olgas) (Fig. 5.31). This group of inselbergs is composed of Mt. Currie conglomerates of probable Cambrian age. These are siliceous, massive, and coarse. Consisting of some thirty rounded domes, Kata Tjuta rises c 460–600 m above the surrounding plains. It is cut into a series of blocks by a near vertical system of joints. It seems that its domes are higher and more resistant than those of Uluru. Its sides have many tafoni, alcoves and

Fig. 5.27 Zhangye Danxia, China. 38°56′36.62″N, 100°5′51.56″E. ©Google Earth. Scale bar = 100 m

flared slopes. Twidale (2010) suggests that like Uluru it is an ancient feature and predates the latest Cretaceous.

The importance of deep clefts is evident in the case of Kata Tjuta, but they are also well shown in the Serra da Capivara in Brazil, where Silurian and Devonian sandstones and conglomerates occur. The area also has multitudes of rock shelters full of important archaeological materials (da Silva Mutzenberg et al. 2015) (Fig. 5.32). Other striking conglomerate landforms occur in the far south of Brazil (de Borba et al. 2024).

5.5 Landscapes Formed in Mudstones and Other Fine-Grained Sedimentary Rocks

5.5.1 Introduction

Mudstones and shales are fine-grained, argillaceous, normally siliciclastic sedimentary rocks whose original constituents were clays or muds. Details of their nature and formation are given in Potter et al. (2005). Mudstones are distinguished from shale by their lack of fissility (parallel layering). They are the most extensive type of sedimentary rock on Earth. They are composed mostly of silicate grains less than 63 μm in diameter. Shale is formed when silt and clay are compressed. It is composed of many thin layers, and it readily splits into thin pieces where these layers meet. Siltstone is composed mostly of silt. It is a form of mudrock with a low clay mineral content. It lacks the laminations and fissility along horizontal lines which are typical of shale. Most of the silt grains are made of quartz and silt is defined as grains that are 2–62 μm in diameter. Many mudstones and shales occur in deposits that have a deep-sea origin, and these are often referred to as *flysch*. These have produced many landslides, as, for example, in the Carpathian Mountains of central Europe. Mudstones can also give rise to badland terrains. In South Africa, the Ecca Shales of the Karoo Group are favoured sites for the formation of aeolian pans (Goudie and Thomas 1985).

5.5.2 Landslips

Stead (2016) provides an excellent review of how shales create slope instability. Shales when encountered in landslides or surface excavations commonly show characteristics such as fissuring, dilational behaviour, negative pore

(a)

(b)

Fig. 5.28 **a** Meteora, Greece. 39°42′55.25″N, 21°37′47.86″E. ©Google Earth. Scale bar = 800 m. **b** Meteora, Greece (Photo by A.S. Goudie)

(a)

(b)

Fig. 5.29 **a** Montserrat, Spain. 41°35′39.38″N, 1°49′57.69″E. ©Google Earth. Scale bar = 500 m. **b** Montserrat, Spain, in 1876 as depicted by Elisée Reclus in his Nouvelle Géographie Universelle, volume 1, p. 818

(a)

(b)

Fig. 5.30 **a** Los Mallos de Riglos, Spain. 42°20′48.34″N, 0°42′56.33″W. ©Google Earth. Scale bar = 200 m. **b** Los Mallos de Riglos, Spain, in 1911. Source: https://commons.wikimedia.org/wiki/Category:Mallos_de_Riglos#/media/File:El_poble_amb_els_Mallos_de_Riglos_al_fons.jpeg. Public Domain

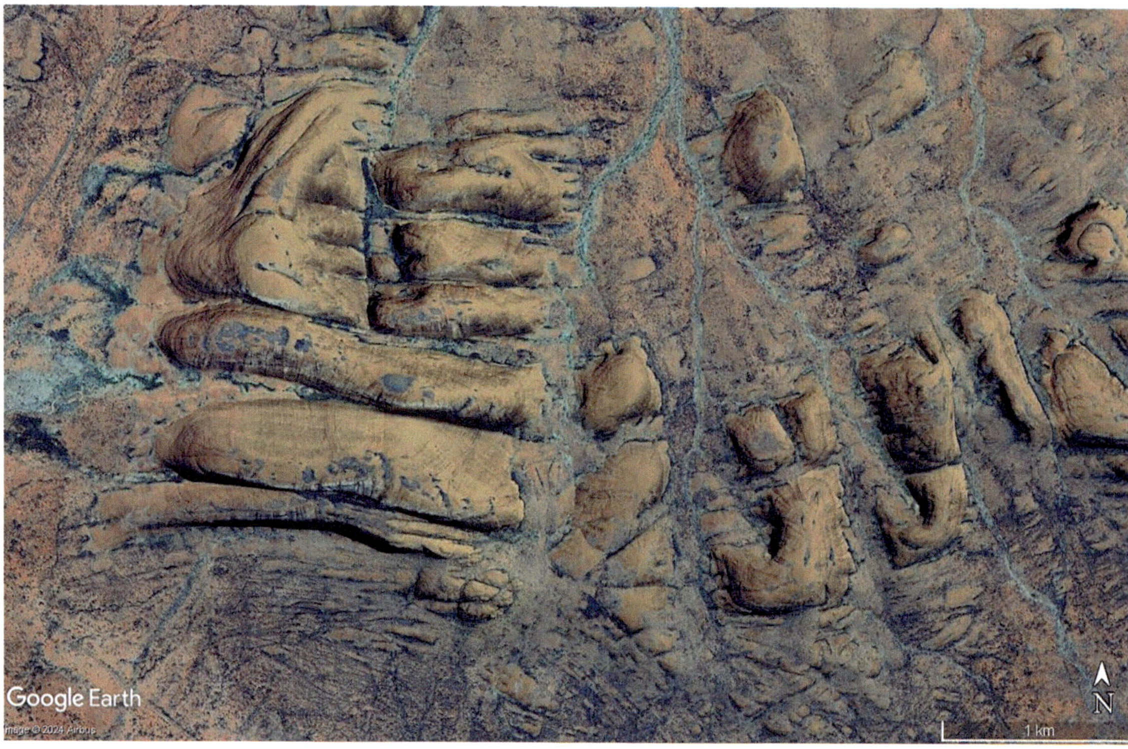

Fig. 5.31 Kata Tjuta (Olgas). 25°17′57.08″S, 130°44′55.58″E. ©Google Earth. Scale bar = 1 km

Fig. 5.32 Serra de Capivara, Brazil. 8°49′6.99″S, 42°32′39.64″W. ©Google Earth. Scale bar = 600 m

water pressures and softening (loss of cohesion). Some shales may have limited cementation (compacted clays) and be highly susceptible to gravitationally induced changes in structure. Other clays may be cemented and the nature of the cement will have an important influence on solution processes and longer term strength changes. Because of their low permeability and anisotropy, groundwater plays a major role in shale slope instability. Their low permeability in comparison to other materials means that pore water pressures may build up either above the shale units, where there is downward flow, or beneath if there are upward seepage gradients; shales thereby vertically compartmentalize groundwater flow. Translational failures in shale slopes are the most commonly reported failure mechanisms. These vary from simple planar failures along weak low shear strength planes to extensive deep-seated multi-block landslides involving shear zones of hundreds of metres in length. Spreading rock slope failures are also commonly observed where major thicknesses of strong denser rock masses overlie weak rock masses. Softening of shale debris material within landslides may lead to extensive earthflow movements.

Landslides in the UK are important in argillaceous rocks, varying from well-documented failures in overconsolidated clays such as London Clay and Gault clays to the landslide-prone slopes of the Lias Formation. The relative impermeability of mudstones means that they frequently throw out springs and develop unstable surfaces across which more permeable overlying sediments may become mobile. This is very evident on the Jurassic Coast in southern England where the Upper Greensand overlies Gault (Cretaceous) clays and Liassic shale beds. At Black Ven, between Charmouth and Lyme Regis, this gives rise to one of the biggest and most active landslide and mudslide complexes in Europe (Arber 1973; Allison 1990; Gallois 2008). The history of their study is given in Goudie and Brunsden (2023) (Fig. 5.33).

A broadly similar situation exists further north in England, along the Cotswold escarpment, where limestones of the Inferior Oolite (Jurassic) overlie Lias clays (Goudie and Parker 1996), causing scarp retreat, landslides, slip troughs, cambering and valley bulging (Goudie 2020). Cambering and valley bulge phenomena occur where competent strata overlie clays and are thought to have been associated with the growth and decay of permafrost (Hollingworth et al. 1944).

Numerous other shale materials have resulted in large landslides in the UK. One of the most famous and well-studied of these is Mam Tor in Derbyshire, where failure involved Carboniferous black Edale (Bowland) shales (Green et al. 2020) under the Mam Tor Beds (a sequence of turbidites of mudstones, siltstones and sandstones) (Fig. 5.34). Mam Tor, the so-called 'shivering mountain', has a landslide that is c 900 m long and over 350 m wide. It is backed by a main scar that is over 80 m high. This scar marks the site of the original rupture, which took place c 3600 years ago, while the uneven ground downslope marks the zone where blocks of rock have slipped and rotated backwards. Yet further downslope, on the toe, there is a hummocky zone where debris flows have occurred. Displacements are ongoing and have caused the closure of a main road (Waltham and Dixon 2000).

In some parts of the world there are quick (or sensitive clays). Quick clay, is also known as Leda clay and Champlain Sea clay. It is found in Canada, Norway, Russia, Sweden, Finland, the United States and other locations. The term quick clay has been used to denote the behaviour of highly sensitive, upraised Quaternary marine clays that, due to post depositional processes, have the tendency to change from a relatively stiff condition to a liquid mass when disturbed. Sensitive clay particles are electrically charged and held together by a thin layer of water molecules, which makes them very unstable and prone to collapse when the water content is disturbed or altered. Quick clay may depart from its stable state due to multiple triggering factors (overloading, erosion, rainfall, slope undercutting, earthquakes, etc.). On failure these marine clays can rapidly mobilise into high velocity flow slides and spreads often completely liquefying in the process (Thakur et al. 2017; Giles 2020).

Some clays are highly fissured and these are especially prone to landslides, having a poor shear strength. They are widespread in Sicily, along the Italian Apennine Mountains, and in many other countries of the Mediterranean area (Rosone et al. 2018). They are also known as scaly or tectonised clays (Cotecchia et al. 2015) and their distinctive fabric commonly forms in response to shearing along a fault zone.

5.5.3 Badlands, USA

The term 'badlands' is probably derived from the French phrase *mauvais terres*. As Archibald Geikie (1882, p. 223), travelling in Wyoming, remarked, 'This expressive name has been given to some of the strangest, and, in many respects, most repulsive scenery in the world. They are tracts of irreclaimable barrenness, blasted and left for ever lifeless and hideous.'

Badlands are indeed barren and are usually intricately dissected by rills and gullies (Howard 2009). Steep-sided residuals rise above gently sloping alluvial or pediment surfaces with interviewing slope angles often reflecting the erosional resistance of different lithologies. Micro-relief is complex, displaying deep desiccation cracks, pipes, rills, sharp divides and vertical faces alternating abruptly with

(a)

(b)

Fig. 5.33 **a** The Black Ven and Spittles landslide complex between Lyme Regis and Charmouth, in Dorset, England. 50°43′59.79″N, 2°55′12.52″W. ©Google Earth. Scale bar = 500 m. A guide to this site is provided by the Charmouth Heritage Coast Centre (2023). **b** Black Ven, Dorset, looking eastwards towards Charmouth. Debris flow in foreground

(a)

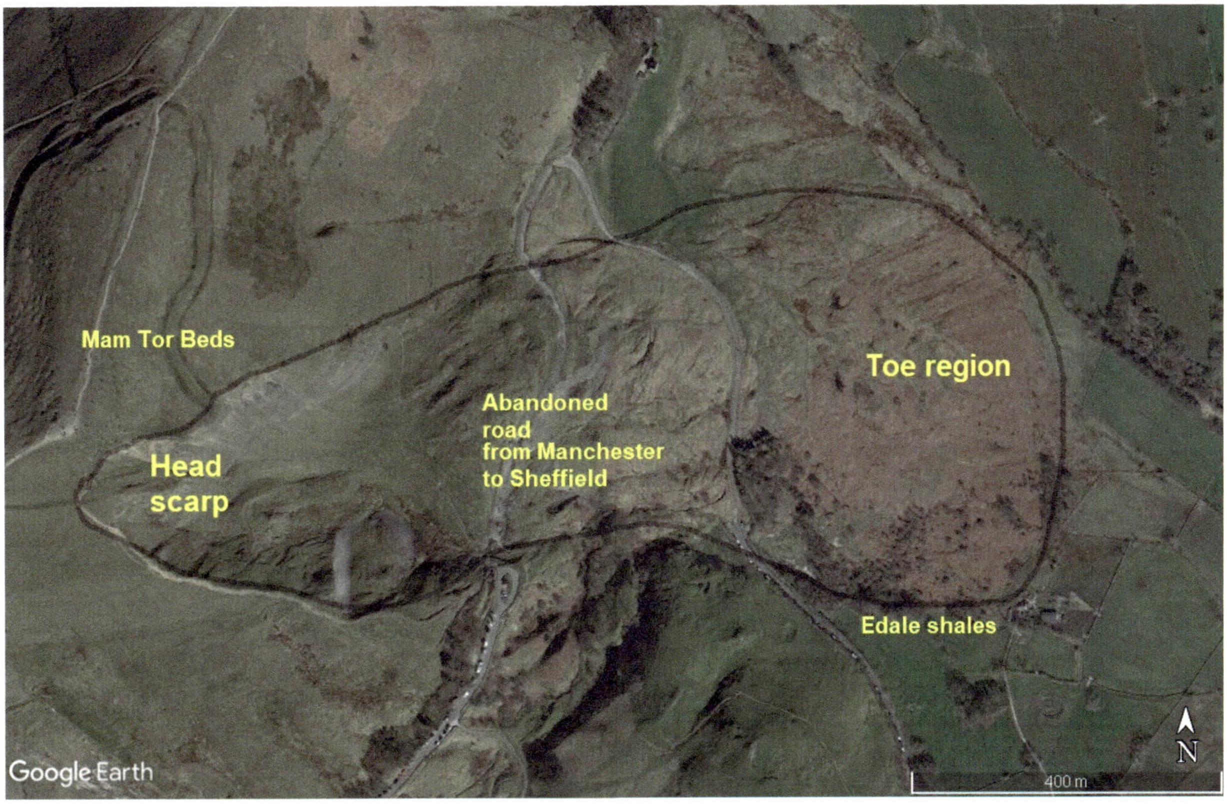

(b)

Fig. 5.34 **a** Mam Tor landlside, Derbyshire, England. 53°20′58.08″N, 1°48′6.38″W. ©Google Earth. Scale bar = 400 m. **b** The abandoned main road at Mam Tor (Photo A.S. Goudie)

rounded forms. Badlands tend to be sites of high rates or erosion and many contribute substantial amounts of sediment to river basins of which they are only a small part (Bryan and Yair 1982). Indeed, erosion may be so rapid that a regolith or soil cannot develop. This does not favour vegetation establishment and, in a positive feedback, the lack of vegetation enhances erosion. Among the processes that mould badlands are mass movements (creep, slab failure and mudflows), fluvial rilling, sub-surface piping (Faulkner 2018), and splash and wash. Also important are frost action, wetting and drying, clay swelling, and slaking. Tectonic uplift may also play a role (Moreno de las Heras and Gallart 2018). Lithology may be a major control of their morphology and distribution. They tend to develop on susceptible lithologies, such as loess, shales and volcanic ash, and where there are soluble cements. Badlands usually have high drainage densities composed of rill and gully systems. They may comprise zones of coalesced hillslope gullies within which little of the pre-gullying terrain remains, as in some of the Mancos Shale Badlands in the USA. Badlands are especially common in areas with at least seasonal drought, in semi-arid and arid areas, Mediterranean and dry-season tropical areas. They may result from natural processes, but their extent may be accentuated by human-induced soil erosion.

Some of the lithological controls of badland development have been disciussed by Campbell (1989). He drew attrention to the importance of bentonite, for bentonitic clay minerals are prone to slaking, cracking and pipe formation. In addition, the association of high percentages of exchangeable sodium with swelling clays causes dispersion. He also provided a Table 8.1 showing the materials, mostly Cretaceous or younger, that make up badlands from many parts of the world.

The badlands of Badlands National Monument (Fig. 5.35) are formed by erosion of the flat-lying clays and silts of the White River Group, Brule and Chadron Formations, of Oligocene age (Gonzalez 2009; Benton et al. 2015). The strata are in most places almost horizontal, with only a slight regional dip of 1–2° to the southeast. They are actively eroding (Schumm 1962). The average yearly precipitation is 406 mm, and vegetation is almost completely absent on the steep badland slopes, though on the adjacent flat areas dense prairie grasslands are common (Engelen 1973). Aspect appears to be an important control of slope forms and processes. South-facing slopes are generally significantly shorter, steeper and straighter in profile than their north-facing counterparts (Churchill 1981). More intense fluvial erosion occurs on the latter and because they receive far less direct solar radiation than south-facing slopes, north-facing slopes retain higher moisture levels and have a greater susceptibility to fluvial erosion and rilling. Moreover, saturation of the regolith, which increases internal pore pressures, enhances their susceptibility to slope failure and mass movements.

The Mancos Shale is of Upper Cretaceous age and is widespread in the American southwest, with onutcrops occuring in Colorado, Utah, Arizona, and New Mexico. It is dominantly composed of mudrock that accumulated in offshore and marine environments of the North American Inland Sea. It was deposited approximately from 95 to 80 mya. It erodes quickly and produces a very intricately dissected landscape (Godfrey et al. 2008) (Fig. 5.36), with drainage densities averaging 60 km km^{-2} (Laronne 1982). In the badlands of Caineville in Wayne County, Utah, USA, the deltaic sandstone of the Emery Member caps the Blue Gate Member of the Mancos Shale, deposited in a sea during the Cretaceous Period. Note the northwest-facing escarpments and the actively eroding badlands (Godfrey 1997; Howard 1997). The Fremont River seems to have cut down by about 65 m since the end of the Bull Lake Pluvial of the Late Pleistocene causing a wave of dissection towards the sandstone cuestas. Howard (2009, Fig. 10.11) provides a map of the area covered by this image and discusses the age and origin of the badlands.

The Toadstool badlands of Nebraska (Fig. 5.37) are so named because of the shape of the eroded rock masses in the area. A hoodoo or toadstool refers to a landform where a large block of erosionally resistant rock overlies more quickly eroding rock underneath, making a toadstool-like feature. The hoodos are made of well-cemented river channel sediment from the Sharps Formation (Oligocene, 30–26 mya) balanced on eroding mudstone of the Scenic Member of the Brule Formation (Oligocene, 33.7–32.0 mya). Other notable toadstool badlands occur in the Bisti area (Fig. 5.38) in predominantly non-marine Upper Cretaceous and Paleocene units along the western and southern margins of the San Juan Basin of northwestern New Mexico (Hutchinson 1981). Lithological differences in flat-lying strata give rise to a series of complex hillslopes.

In Arizona lies the well-named Painted Desert (Fig. 5.39). Here the badlands have developed in the sedimentary rocks of the Upper Triassic Chinle Group – siltstone, mudstone and shale. It contains some mesas that are capped by resistant, lacustrine limestone layers and volcanic flows. The area is also noted for its petrified forests (Ash and Beus 1987). Details of the stratigraphy of the area are provided in https://www.nps.gov/pefo/learn/nature/geologicformations.htm (Accessed 9th May 2024).

The gullied badlands at Zabriskie Point in Death Valley, California, are composed of highly erodible Pliocene lake sediments (Fig. 5.40). These were deposited in Furnace Creek Palaeolake, which formed about 9 mya. During several million years of its history, saline muds, gravels from nearby highlands and ash falls from the then-active Black Mountain volcanic field accumulated at the lake bottom

Fig. 5.35 **a** The Badlands of South Dakota, USA. 43°51′36.87″N, 102°18′25.73″W. ©Google Earth. Scale bar = 200 m. **b** The badlands of South Dakota (Photo by A.S. Goudie)

(a)

(b)

Fig. 5.36 **a** Mancos Shale Badlands, Utah, USA. 38°16′18.13″N, 110°55′22.33″W. ©Google Earth. Scale bar = 7 km. **b** Mancos Shale badland in Colorado (Photo A.S. Goudie). **c** Shale badlands at Caineville, Utah, USA. 38°22′19.72″N, 110°58′31.20″W. ©Google Earth. Scale bar = 2 km

(c)

Fig. 5.36 (continued)

and tephrochronologic and paleomagnetic data indicate that the upper member of the Furnace Creek Formation dates to between c 4.1 and c 3.2 mya (Knott et al. 2016).

5.5.4 Badlands—Spain and Italy

In the Ebro Basin of Navarre in northern Spain, there is a remarkable area of badlands called Bardenas Reales. Here the mean annual rainfall is only 300 mm, and the local lithologies consists of various erodible Tertiary beds. Of these most important are (i) the Lerín Gypsum Formation, (ii) the Tudela Formation and (iii) limestones of the Sancho Abarca Formation. The Tudela Formation, which is of Lower–Middle Miocene age, is a sequence of about 320 m of multi-coloured clays interbedded with lacustrine limestone, sand and gypsum. These are probably the distal facies of alluvial fans that once came from the Pyrenees. The steep slopes and the very sparse vegetation have allowed the development of an imposing badlands landscape (Desir and Marín 2007; 2013). The centre of the Bardenas Reales depression is occupied by Holocene clays and silts that come from the washing out of the clays of the Tudela Formation. This Holocene material, which has low cohesion, is also susceptible to erosion so that deep gullies have developed in it as well (Fig. 5.41).

Badlands are also well known from the semi-arid Almeria district of south eastern Spain, where the Tabernas Badlands occur in marly gypsum-calcareous mudstones of Upper Miocene age (Calvo-Cases et al. 2014).

Badlands also occur in various parts of Italy, including Tuscany. The typical landforms on clayey slopes are represented by *calanchi,* composed of systems of rills and gullies separated by sharp and steep ridges, and by small, bare domes called *biancane* (del Monte 2017). They have developed on uplifted Plio-Pleistocene marine sediments composed of alternating beds of clay, marl-clay, silt–clay and silt (Anselmi et al. 1994). An example from the Calanchi di Atri in Abruzzo is shown in Fig. 5.42. These badlands, are locally known as 'Dantesque pits of hell' or 'devil's scratches' (Moretti and Rodolfi 2000). Maps of both types of badland are provided by Alexander (1982).

5.6 Silcrete Duricrusts

The word 'duricrust' was originally introduced by Woolnough (1927) who subsequently defined the term thus (Woolnough 1930, pp. 124–5):

> The widespread chemically formed capping in Australia, resting on a thoroughly leached sub-stratum… The nature of the

Fig. 5.37 Toadstool Badlands, Nebraska, USA. 42°51′14.96″N, 103°35′10.38″W. ©Google Earth. Scale bar = 500 m

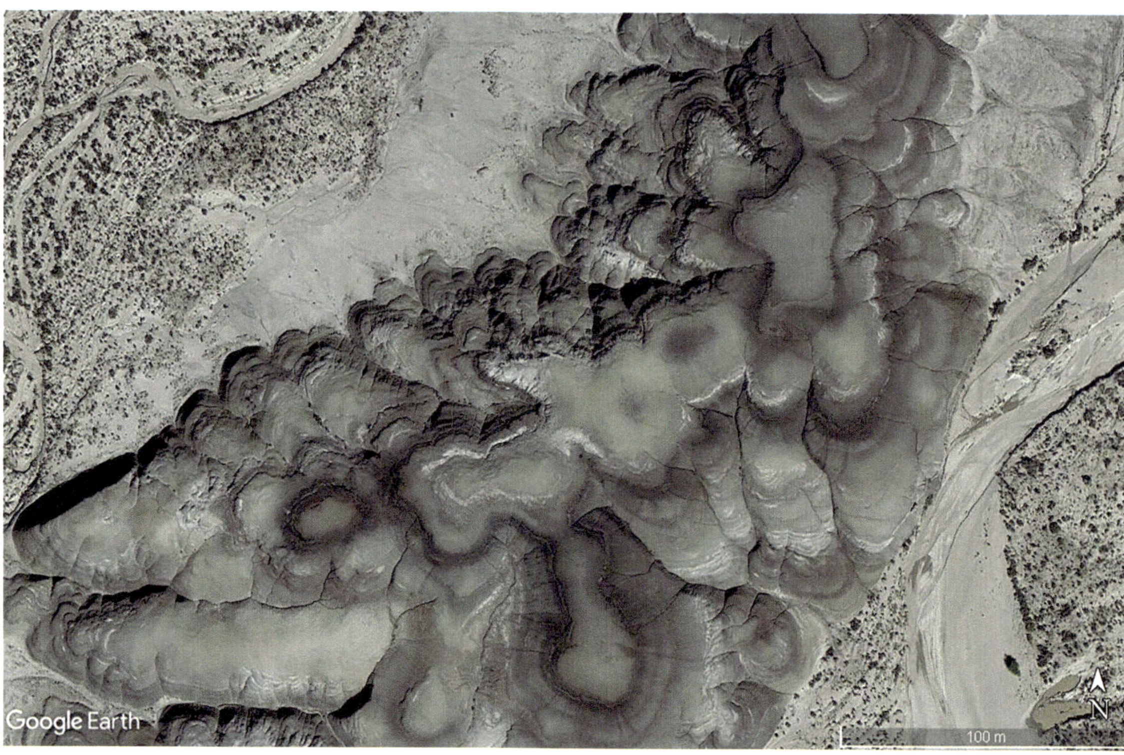

Fig. 5.38 Bisti Badlands, New Nexico, USA. 36°16′38.05″N, 108°9′25.00″W. ©Google Earth. Scale bar = 100 m

Fig. 5.39 The Painted Desert, Arizona, USA. 35°10′17.78″N, 109°49′56.25″W. ©Google Earth. Scale bar = 200 m

> deposit varies from a mere infiltration of pre-existing surface rock, to a thick mass of relatively pure chemical precipitate…

The mineral matter deposited from solution falls into three main groups:

(a) Aluminous and ferruginous
(b) Siliceous
(c) Calcareous and magnesian

He believed that bedrock was an important influence on the distribution of these three types, which in effect are broadly equivalent to (a) laterites, bauxites, and ferricretes, (b) silcretes, and (c) calcretes and dolocretes.

Goudie (1973, p. 5) proposed a definition which resulted from a synthesis of a various definitions that had already been developed for the individual types, and stressed their essentially sub-aerial and near-surface origin and nature:

> A product of terrestrial processes within the zone of weathering in which either iron and aluminium sesquioxides (in the case of ferricretes and alcretes) or silica (in the case of silcrete) or calcium carbonate (in the case of calcrete) or other compounds (in the case of magnesicrete and the like) have dominantly accumulated in and/or replaced a pre-existing soil, rock, or weathered material, to give a substance which may ultimately develop into an indurated mass.

Siliceous duricrusts (silcretes) are known from various parts of the world. They form when silica is dissolved and resolidifies as a cement. Silcrete is a hard and resistant material, and although different in origin and nature, appears similar to quartzite (Taylor and Eggleton 2017). Summerfield (1982) discussed those in the Kalahari and elsewhere in southern Africa, as at Grahamstown. However, they are even more extensive in Australia (Langford-Smith 1978), especially in the Lake Eyre Basin, where a strking residual rises up after the salty lake floor (Fig. 5.43).

The Kanku-Breakaways Conservation Park is a protected area, also in South Australia, 33 km north of Coober Pedy (Fig. 5.44). The geology and eroding plateaux for which the area is noted, formed from rocks deposited about 115 mya. Uplift of the basin and withdrawal of the Cretaceous sea exposed the rocks to around 60 million years of weathering and erosion, causing bleaching and iron-staining which has given them their spectacularly varied colours. Alternating wet and dry climatic conditions during this period allowed silica to accumulate in the soil profile, eventually cementing the near-surface as hard silcrete. Resistant to erosion, silcrete-capped areas remain, creating striking flat-topped mesas. Erosion has carved steep-sided escarpments (often called breakaways in Australia) and the silcrete fragments

(a)

(b)

Fig. 5.40 **a** Heavily-gullied badlands at Zabriskie Point, Death Valley, California, USA. 36°25′N, 116°48′W. Scale bar = 300 m. **b** Zabriskie Point, California, USA (Photo by A.S. Goudie)

Fig. 5.41 Badlands at Badenas Reales, Spain. 42°13′16.93″N, 1°26′31.90″W. ©Google Earth. Scale bar = 100 m

Fig. 5.42 The Calanchi di Atri, Italy. 42°34′21.74″N, 13°57′43.82″E. ©Google Earth. Scale bar = 300 m

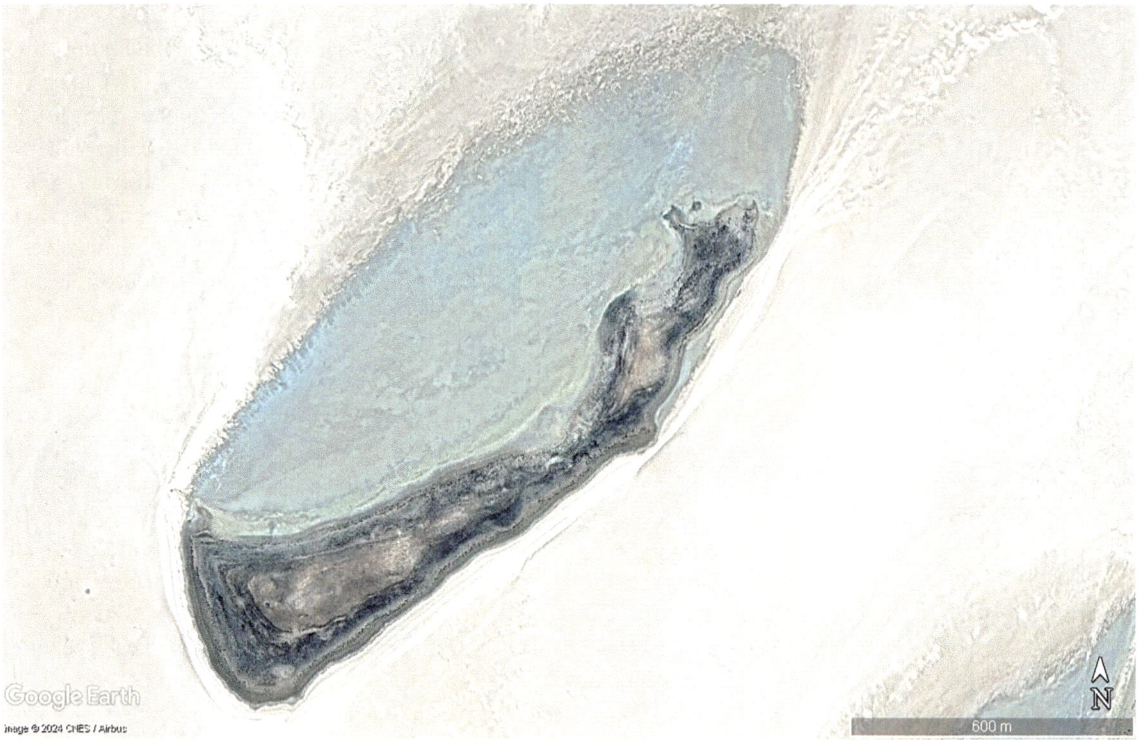

Fig. 5.43 Silcrete Island, South Australia. This is a residual of a silcrete sheet located within Lake Eyre (Kati Thanda). 28°52′44.17″S, 137°6′9.90″E. ©Google Earth. Scale bar = 600 m

Fig. 5.44 Kanku Breakaways, silcrete mesa, South Australia. 28°50′S, 134°, 42′E. ©Google Earth. Scale bar = 200 m

Fig. 5.45 Miracking silcreted palaeochannel, Australia. 27°55′10.64″S, 134°39′31.14″E. ©Google Earth. Scale bar = 20 km

Fig. 5.46 The Valley of Stones—a sarsen valley train—in Dorset, southern England. 50°41′3.32″N, 2°34′22.61″W. ©Google Earth. Scale bar = 50 m. Note the prehistoric stone circle

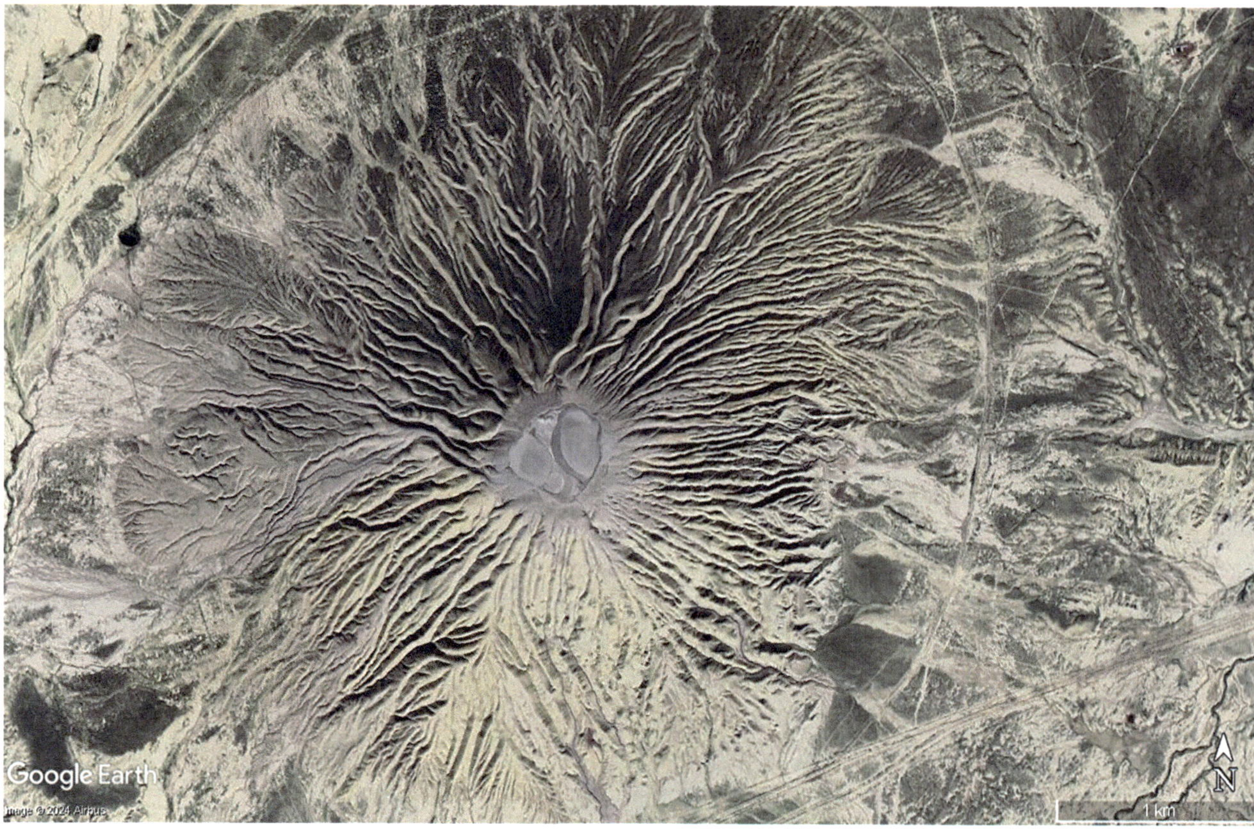

Fig. 5.47 Mud volcano, Azerbaijan. 40°9′57.93″N, 49°18′1.29″E. ©Google Earth. Scale bar = 1 km

(*gibbers*) can be removed over long distances from the foot of these eminences. Not all Australian breakaways are capped by silcretes. Laterite-capped examples are also widespread (see, for example, Mulcahy (1967) on south-western Australia).

Also in the northern part of South Australia is another intriguing feature. This is the ancient Mirackina palaeoriver (Fig. 5.45). It is an instance of relief inversion and is the result of silicification of gravels having occurred along a former river channel. The channel fill comprises fluvial conglomerate and sandstone; has a siliceous and ferruginous duricrust caprock; and overlies kaolinised and weakly silicified Bulldog Shale. This duricrust protected the channel sediments from subsequent erosion and so they were left upstanding as a 200 km long sinuous ridge (Pain and Ollier 1995; Williams et al. 2013). The palaeochannel sediments are thought to be Miocene in age, though their silicification may have been completed later (possibly during the Pliocene) (McNally and Wilson 1995).

In the mid-Cenozoic, silcretes (called sarsen stones or puddingstones in southern England) developed in large tracts of western Europe, including the Paris Basin. They are particularly notable for forming lines of boulders along dry valleys, and many were probably valley-train silcretes that developed under warm conditions in the Eocene and which in the cold phases of the Pleistocene were broken up by frost and moved by solifluxion (Migoń 2020a, b) (Fig. 5.46).

5.7 Mud Volcanoes

Mud volcanoes are not proper volcanoes, in the sense that they do not involve the extrusion of magma. They are features formed by the periodic venting of fluid mud, water and hydrocarbons, which creates mounds that may be up to 3500 m in diameter and 100 m in height. Near their summits there may be cones and pools. They tend to form in compressional tectonic settings such as on convergent plate margins. *Inter alia* they are known from Azerbaijan (Fig. 5.47) (Aliyev et al. 2024) and the northern Appenines of Italy (Castaldini and Coratza 2017) (Fig. 5.48). The Azerbaijan examples are numerous (c 350–400 in the South Caspian Basin area) and amongst the biggest on Earth. The high concentration of mud volcanoes in Azerbaijan is due to the country's location on the boundary between the Eurasian and Arabian tectonic plates. The collision between these two plates has created a zone of intense tectonic activity, which has led to the formation of mud volcanoes and other geological features in the region. The mud volcanoes

(a)

(b)

Fig. 5.48 **a** Mud volcano at Salse di Nirano, near Modena, Italy. 44°30′48.89″N, 10°49′25.32″E. ©Google Earth. Scale bar = 50 m. **b** Mud volcanoes in the Appenines near Modena, Italy, (Photo A.S. Goudie)

have a range of morphologies, including cones, shields and domes, and these are clearly visible on Google Earth (Komatsu and Feyzullayev 2024).

The mud volcanoes of Italy are clustered in three main groups: the northern Apennines (Pede-Apennines margin of Emilia-Romagna), the central Apennines (eastern Marche and Abruzzo) and Sicily.

The Pede-Apennine region between, Parma and Bologna, has nearly 30% of all those found in Italy. The clayey materials ejected from the craters cover the surrounding ground with mudflows, and because of their their fluidity these mudflows can cover distances of up to 100 m from the vent.

5.8 Conclusions

Landscapes composed of the various types of sedimentary rocks cover a very substantial parts of Earth's land surface. Sandstones, conglomerates and fine-grained mudstones, clays and shales, have distinctive geomorphological impacts, though they may also have similarities. The landscapes associated with another type of sedimentary rock—limestones—are discussed in the next chapter.

References

Aghassy J (1970) Jointing, drainage, and slopes in a West African epeirogenetic savanna landscape. An Ass Am Geogr 60(2):286–298

Alexander D (1982) Difference between "calanchi" and "biancane) Badlands in Italy. In: Bryan R, Yair A (eds) Badland geomorphology and piping. GeoBooks, Norwich, pp 71–87

Aliyev A, Huseynov D, Abbasov O, Rashidov T, Kangarli I (2024) Mud volcanoes of Azerbaijan: the unique natural objects of the geoheritage. Geoheritage 16(1). https://doi.org/10.1007/s12371-024-00931-3

Allison RJ (ed) (1990) Landslides of the dorset coast. British Geomorphological Research Group Field Guide, p 127

Anselmi B, Crovato C, D'Angelo L, Grauso S (1994) I Calanchi Di Atri (Abruzzo): Caratteri mineralogici, geotecnici e geomorfologici. Quaternario 7:145–158

Arber MA (1973) Landslips near Lyme Regis. Proc Geol Ass 84(2):121-I33

Ash SR, Beus SS (1987) Petrified forest national park, Arizona. Geol Soc Am Centennial Field Guide—Rocky Mountain Section, pp 405–410

Avigad D, Sandler A, Kolodner K, Stern RJ, McWilliams M, Miller N, Beyth M (2005) Mass-production of Cambro-Ordovician quartz-rich sandstone as a consequence of chemical weathering of Pan-African terranes: environmental implications. Earth Planet Sci Lett 240(3–4):818–826

Baars DL (2000) The colorado plateau. A geologic history. Univ New Mexico Press, Albuquerque, p 254

Barnes FA (1987) Canyon Country arches and bridges and other natural rock openings. Canyon Country Publications, Moab, p 416

Benton RC, Terry DO, Evanoff E, McDonald HG (2015) The white river badlands: geology and paleontology. Indiana Univ Press, Bloomington and Indianapolis, p 240

Blair RW (1987) Development of natural sandstone arches in south-eastern Utah. In: Gardiner V (ed) International geomorphology 1986, part 2. Wiley, Chichester, pp 597–604

Bourman RP, Ollier CD, Buckman S (2015) Inselbergs and monoliths: a comparative review of two iconic Australian landforms, Uluru (Ayers Rock) and Burringurrah (Mount Augustus). Z Geomorph NF 59(2):197–227

Bowden DJ (1980) Sub-laterite cave systems and other pseudokarst phenomena in the humid tropics: the example of the Kasewe Hills, Sierra Leone; sub-laterite cave systems and other pseudokarst phenomena in the humid tropics: the example of the Kasewe Hills, Sierra Leone. Z Geomorph 24(1):77–90

Bradley WC (1963) Large-scale exfoliation in massive sandstones of the Colorado Plateau. Geol Soc Am Bull 74(5):519–528

Brierley G et al (2011) Naming conventions in geomorphology: contributions and controversies in the sandstone landscape of Zhangjiajie Geopark, China. Earth Surf Proc Landf 36(14):1981–1984

Buckman S, Morris RH, Bourman RP (2021) Fire-induced rock spalling as a mechanism of weathering responsible for flared slope and inselberg development. Nat Commun 12(1):1–14

Busche D (2001) Early Quaternary landslides of the Sahara and their significance for geomorphic and climatic history. J Arid Environ 49(3):429–448

Busche D (2010) North-eastern Niger: sandstone landscape of the Sahara. In: Migoń P (ed) Geomorphological landscapes of the world. Springer, Dordrecht, pp 171–182

Calvo-Cases A et al (2014) Badlands in the tabernas basin, betic chain. In: Gutiérrez F, Gutiérrez M (eds) Landscapes and landforms of Spain. Springer, Dordrecht, pp 197–211

Campbell IA (1989) Badlands and badland gullies. In: Thomas DSG (ed) Arid zone geomorphology. Belhaven Press, London, pp 159–183

Castaldini D, Coratza P (2017) Mud volcanoes in the Emilia-Romagna Apennines: small landforms of outstanding scenic and scientific value. In: Soldati M, Marchetti M (eds) Landscapes and landforms of Italy. Springer, Cham, pp 225–234

Chardon D, Chevillotte V, Beauvais A, Grandin G, Boulangé B (2006) Planation, bauxites and epeirogeny: one or two paleosurfaces on the West African margin? Geomorphology 82(3–4):273–282

Charmouth Heritage Coast Centre (2023) Charmouth, Black Ven and the Spittles. Charmouth, p 24

Churchill RR (1981) Aspect-related differences in badlands slope morphology. Ann Ass Am Geogr 71(3):374–388

Cotecchia F, Vitone C, Santaloia F, Pedone G, Bottiglieri O (2015) Slope instability processes in intensely fissured clays: case histories in the Southern Apennines. Landslides 12:877–893

Cremaschi M, Zerboni A, Spötl C, Felletti F (2010) The calcareous tufa in the Tadrart Acacus Mt. (SW Fezzan, Libya): an early Holocene palaeoclimate archive in the central Sahara. Palaeogeogr Palaeoclimatol Palaeoecol 287(1–4):81–94

da Silva Mutzenberg D, de Barros Correa AC, de Azevêdo Cavalcanti Tavares B, Cisneiros D (2015) Serra da Capivara National Park: Ruinform landscapes on the Parnaiba cuesta. In: Veira BC, Salgado A, Santos L (eds) Landscapes and landforms of Brazil. Springer, Dordrecht, pp 253–263

de Borba AW, Sccoti AAV, Guadagnin F (2024) Conglomerate geomorphology in the Precambrian Camaquã basin of southernmost Brazil: contribution to a global panorama. J S Am Earth Sci 104820

del Monte M (2017) The typical badlands landscapes between the Tyrrhenian Sea and the Tiber River. In: Soldati M, Marchetti M (eds) Landscapes and landforms of Italy. Springer, Cham, pp 281–291
Desir G, Marín C (2007) Factors controlling the erosion rates in a semi-arid zone (Bardenas Reales, NE Spain). CATENA 71(1):31–40
Desir G, Marín C (2013) Role of erosion processes on the morphogenesis of a semiarid badland area. Bardenas Reales (NE Spain). Catena 106:83–92
Dixon JC (2010) Canyonlands and arches: windows on landscapes in the American Southwest. In: Migoń P (ed) Geomorphological landscapes of the world. Springer, Dordrecht, pp 39–47
Doerr SH (1999) Karst-like landforms and hydrology in quartzites of the Venezuelan Guyana shield: pseudokarst or "real" karst? Z Geomorph 43(1):1–17
Duszyński F, Migoń P (2020a) Brimham Rocks—Sandstone ruiniform relief in the North of England. In: Goudie A, Migoń P (eds) Landscapes and landforms of England and Wales. Springer, Dordrecht, pp 437–451
Duszyński F, Migoń P (2020b) Sandstone landforms of the High Weald. In: Goudie A, Migoń P (eds) Landscapes and landforms of England and Wales. Springer, Dordrecht, pp 103–118
Embabi NS (ed) (2018) Landscapes and landforms of Egypt. Springer, Cham, p 336
Engelen GB (1973) Runoff processes and slope development in Badlands National Monument, South Dakota. J Hydrol 18(1):55–79
Eramo G, Muntoni IM, Aprile A, Pallara M, Rotunno R, Zerboni A, di Lernia S (2020) Networking through pottery characterisation at Takarkori rock shelter (Libyan Sahara, 10,200–4650 cal BP). Archaeol Anthropol Sci 12(9):1–33
Faulkner H (2018) The role of piping in the development of badlands. In: Nadal-Romero E, Murillo JFM, Kuhn NJ (eds) Badlands dynamics in a context of global change. Elsevier, Amsterdam, pp 191–216
Gallois RW (2008) Geological controls on the failure mechanisms within the Black Ven-Spittles landslip complex, Lyme Regis, Dorset. Proc Ussher Soc 12(1):9–14
Geikie A (1882) Geological sketches at home and abroad. Macmillan, London, p 382
Giles DP (2020) Quick clay behaviour in sensitive Quaternary marine clays—a UK perspective. Geol Soc Lon Eng Geol Spec Pub 29(1):205–221
Godfrey AE (1997) Mass movement of Mancos Shale crust near Caineville, Utah: a 30-year record. Geogr Ann: Series A 79(3):185–194
Godfrey AE, Everitt BL, Duque JFM (2008) Episodic sediment delivery and landscape connectivity in the Mancos Shale badlands and Fremont River system, Utah, USA. Geomorphology 102(2):242–251
Gonzalez MA (2009) Badlands of the Northern Great Plains: hell with the fires out. In: Migoń P (ed) Geomorphological landscapes of the world. Springer, Dordrecht, pp 29–38
Goudie AS (1973) Duricrusts in tropical and subtropical landscapes. Clarendon Press, Oxford, p 174
Goudie AS (2007) Mega-yardangs: a global analysis. Geogr Compass 1(1):65–81
Goudie AS (2008) Wheel across the desert. Silphium Press, London, p 205
Goudie AS (2020) The Cotswolds. In: Goudie A, Migoń P (eds) Landscapes and landforms of England and Wales. Springer, Cham, pp 297–315
Goudie AS, Piggott NR (1981) Quartzite tors, stone stripes and slopes at the Stiperstones, Shropshire, England. Biul Peryglac 28:47–56
Goudie AS, Thomas DSG (1985) Pans in southern Africa with particular reference to South Africa and Zimbabwe. Z Geomorph 29(1):1–9
Goudie A, Gardner R (1992) Stiperstones: the Devil's tors. Discovering landscape in England & Wales. Chapman and Hall, London, pp 68–69
Goudie A, Parker A (1996) The geomorphology of the Cotswolds. Cotteswold Naturalists' Field Club, Cheltenham, p 139
Goudie AS, Brunsden D (2023) Geological pioneers of the Jurassic Coast. Oxford Univ Press, Oxford, p 191
Goudie A, Migon P, Allison R, Rosser N (2002) Sandstone geomorphology of the Al Quwayra area of south Jordan. Z Geomorph 46:365–490
Grab SW, Goudie AS, Viles HA, Webb N (2011) Sandstone geomorphology of the Golden Gate Highlands National Park, South Africa, in a global context. Koedoe 53(1):1–14
Green S, Rutter E, Arkwright C (2020) The Mam Tor creeping landslide. In: Goudie A, Migoń P (eds) Landscapes and landforms of England and Wales. Springer, Cham, pp 417–435
Guerrero J, Gutiérrez F, Gutiérrez M (2014) Conglomerate monoliths and karst in the Ebro Cenozoic Basin, NE Spain. In: Gutierrez M (ed) Gutiérrez. Landscapes and landforms of Spain. Springer, Dordrecht, pp 81–89
Gutiérrez F, Gutiérrez M, Gracia FJ (2005) Karst, neotectonics and periglacial features in the Iberian Ranges. In: Field trip C-5. Sixth international conference on geomorphology, Zaragoza, p 58
Härtel H, Cílek V, Jackson A, Herben T, Williams R (eds) (2007) Sandstone landscapes. Academia, Praha, p 493
Hollingworth SE, Taylor JH, Kellaway GA (1944) Large-scale superficial structures in the Northampton Ironstone field. QJ Geol Soc 100(1–4):1–44
Howard AD (1997) Badland morphology and evolution: Interpretation using a simulation model. Earth Surf Proc Landf 22(3):211–227
Howard AD (2009) Badlands and gullying. In: Parsons AJ, Abrahams AD (eds) Geomorphology of desert environments. Springer, Cham, pp 265–299
Howard AD, Kochel RC, Holt HE (eds) (1988) Sapping features of the Colorado Plateau: a comparative planetary geology field guide. Scientific and Technical Information Division, NASA, Washington DC, p 108
Hutchinson PJ (1981) Stratigraphy and paleontology of Bisti Badlands, San Juan County, New Mexico. AAPG Bull 65(3):562
Janeras M et al (2017) Multi-technique approach to rockfall monitoring in the Montserrat massif (Catalonia, NE Spain). Eng Geol 219:4–20
Knight J (2024) Landforms and landscapes of the Ennedi Plateau. In: Knight J, Marlo S, Zerboni A (eds) Landscapes and landforms of the Central Sahara. Springer, Cham, pp 41–48
Knott JR, Machette MN, Liddicoat WE, Sarna-Wojiciki AM (2016) Stop 5: the ancient Furnace Creek Basin at Zabriskie Point. U.S. Geol Survey Open-File Report 2016–1064:22–30
Komatsu G, Feyzullayev AA (2024) Geomorphology of subaerial mud volcanoes in Azerbaijan: issues about edifice construction and degradation. Geomorphology 463:109352
Laity JE, Malin MC (1985) Sapping processes and the development of theater-headed valley networks on the Colorado Plateau. Geol Soc Am Bull 96(2):203–217
Langford-Smith T (ed) (1978) Silcrete in Australia. Department of Geography, University of New England, Armidale, p 304
Laronne J (1982) Sediment and solute yield from Mancos Shale hillslopes. In: Bryan R, Yair A (eds) Badland geomorphology and piping. GeoBooks, Norwich, pp 181–193
Le Heron DP (2018) An exhumed Paleozoic glacial landscape in Chad. Geology 46(1):91–94

Leeder M (1991) Sedimentology and sedimentary basins. From turbulence to tectonics. Blackwell Science, Oxford, p 592
Liao R, Chen W, Liu P, Wang L, Xia Q, Yang X, He Y, Guo Q (2024) Mineral transformation, element transport and hydrological impact in weathering at the Bingling Temple Grottoes: implications for weathering in alkaline environments in NW China. Catena 239: 107966
Magalhães Junior AP, de Paula Barros LF, Felippe MF (2015) Southern Serra do Espinhaço: the impressive plateau of quartzite ridges. In: Vieira B, Salgado A, Santos L (eds) Landscapes and landforms of Brazil. Springer, Dordrecht, pp 359–379
Mainguet M (1972) Le modelé des grès. Problèmes généraux. Institut Géogr National, Paris, p 657
Maxwell TA (1980) Geomorphology of the Gilf Kebir. Geogr J 146:76–83
McCauley JF, Breed CS, Grolier MJ (1982a) The interplay of fluvial, mass-wasting and eolian processes in the eastern Gilf Kebir region. In: El-Baz F, Maxwell TA (eds) Desert landforms of southwest Egypt: a basis for comparison with Mars. NASA, Washington DC, pp 207–239
McCauley JF, Schaber GG, Breed CS, Grolier MJ, Haynes CV, Issawi B, Elachi C, Blom R (1982b) Subsurface valleys and geoarcheology of the eastern Sahara revealed by shuttle radar. Science 218:1004–1020
McNally GH, Wilson IR (1995) Silcretes of the Mirackina Palaeochannel, Arckaringa, South Australia. J Austral Geol Geophys 16(3):295–302
Michel P (1969) Morphogenesis and pedogenesis. African Soils 14:109–141
Migoń P (2020a) Geomorphology of conglomerate terrains—global overview. Earth Sci Rev 208:103302
Migoń P (2020b) Sarsens—the maker of upland scenery of southern England: from Mid-Cenozoic gravel plains to Neolithic landscapes. In: Goudie AS, Migoń P (eds) Landscapes and landforms of England and Wales. Springer, Cham, pp 317–329
Migoń P, Goudie A (2014) Sandstone geomorphology of south-west Jordan, Middle East. Quaest Geogr 33(3):123–130
Migoń P, Duszyński F, Goudie AS (2017) Rock cities and ruiniform relief: forms–processes–terminology. Earth Sci Rev 171:78–104
Moreno-de las Heras M, Gallart F (2018) The origin of badlands. In: Nadal-Romero E, Martinez-Murillo M, Kuhn NJ (eds) Badlands dynamics in a context of global change. Elsevier, Amsterdam, pp 27–59
Moretti S, Rodolfi G (2000) A typical "calanchi" landscape on the Eastern Apennine margin (Atri, Central Italy): geomorphological features and evolution. CATENA 40(2):217–228
Morris TG, Ritter SM, Laycok DP (2012) Geology unfolded. BYU Press, Provo, p 72
Mulcahy MJ (1967) Landscapes, laterites and soils in southwestern Australia. In: Jennings JN, Mabbutt JA (eds) Landform studies from Australia and New Zealand. Cambridge Univ Press, Cambridge, pp 211–230
Munthe J (2002) Arches of the Escalante Canyons and Kaipoarowits Plateau. Arch Hunter Books, Thompson Springs, p 192
Netoff DI, Shroba RR (1993) Morphology and possible origin of giant weathering pits in the Entrada Sandstone, southeastern Utah: preliminary findings. US Geol Surv, Open File Report 93-390. Denver, Colorado, p 45
Ori GG, Roveri M (1987) Geometries of Gilbert-type deltas and large channels in the Meteora Conglomerate, Meso-Hellenic basin (Oligo-Miocene), central Greece. Sedimentology 34(5):845–859
Osborn G, Duford JM (1981) Geomorphological processes in the inselberg region of south-western Jordan. Palest Explor Q 113(1):1–17
Pain CF, Ollier CD (1995) Regolith stratigraphy: principles and problems. J Austral Geol Geophys 16(3):197–202
Peel RF (1941) Denudation landforms of the central Libyan Desert. J Geomorph 4:3–23
Peifer D, Persano C, Hurst MD, Bishop P, Fabel D (2021) Growing topography due to contrasting rock types in a tectonically dead landscape, Earth Surf. Dynam 9:167–181
Peng H (2020) China Danxia. Springer Nature, Singapore, p 391
Potter PE, Maynard JB, Depetris PJ (2005) Mud and mudstones. Springer, Heidelberg, p 297
Rassios AE, Ghikas D, Dilek Y, Vamvaka A, Batsi A, Koutsovitis P (2020) Meteora: a billion years of geological history in Greece to create a World Heritage Site. Geoheritage 12(4):83
Reading HG (ed) (1978) Sedimentary environments and facies. Blackwell Scientific, Oxford, p 557
Rosone M, Ziccarelli M, Ferrari A, Farulla CA (2018) On the reactivation of a large landslide induced by rainfall in highly fissured clays. Eng Geol 235:20–38
Sauro F et al (2019) Genesis of giant sinkholes and caves in the quartz sandstone of Sarisariñama tepui, Venezuela. Geomorphology 342: 223–238
Schoeneich P, Bouzou I (1996) Landslides in Niger (Western Africa). In: Senesset K (ed) Landslides. Balkema, Rotterdam, pp 1967–1972
Schumm SA (1962) Erosion on miniature pediments in Badlands National Monument, South Dakota. Geol Soc Am Bull 73(6):719–724
Schwarz T (1994) Ferricrete formation and relief inversion: an example from Central Sudan. CATENA 21(2–3):257–268
Stead D (2016) The influence of shales on slope instability. Rock Mech Rock Eng 49:635–651
Summerfield MA (1982) Distribution, nature and probable genesis of silcrete in arid and semi-arid southern Africa. Catena Suppl 1:37–65
Taylor G, Eggleton RA (2017) Silcrete: an Australian perspective. Austral J Earth Sci 64(8):987–1016
Thakur V, L'Heureux J-S, Locat A (eds) (2017) Landslides in sensitive clays. Springer, Cham, p 603
Thomas C, Willis G (2019) Landscape Arch, Delicate Arch, and Double Arch in Arches National Park, Southeastern Utah. Utah Geol Ass Publ 1:1–12
Tricart J (1972) Landforms of the humid tropics, forests and savannas. Longman, London, p 306
Twidale CR (2010) Uluru (Ayers Rock) and Kata Tjuta (The Olgas): inselbergs of central Australia. In: Migoń P (ed) Geomorphological landscapes of the world. Springer, Dordrecht, pp 321–332
Viles HA, Goudie AS (2004) Biofilms and case hardening on sandstones from Al-Quwayra, Jordan. Earth Surf Proc Landf 29(12):1473–1485
Waltham T (1994) The sandstone fantasy of Petra. Geol Today 10(3):105–111
Waltham AC, Dixon N (2000) Movement of the Mam Tor landslide, Derbyshire, UK. QJ Eng Geol Hydrogeol 33(2):105–123
Williams RM, Irwin RP, Burr DM, Harrison T, McClelland P (2013) Variability in Martian sinuous ridge form: case study of Aeolis Serpens in the Aeolis Dorsa, Mars, and insight from the Mirackina paleoriver, South Australia. Icarus 225(1):308–324
Woolnough WG (1927) The duricrust of Australia. Jour Proc Royal Soc NSW 61:1–53
Woolnough WG (1930) The influence of climate and topography in the formation and distribution of products of weathering. Geol Mag 67(3):123–132
Wray RA (2010) The Gran Sabana: the world's finest quartzite karst? In: MIgoń P (ed) Geomorphological landscapes of the world. Springer, Dordrecht, pp 79–88
Yakubu M, Ojanuga AG (2013) Pedogenesis, weathering status and mineralogy of the soils on ironstone plateaux (laterites), Sokoto Nigeria. Bayero J Pure Appl Sci 6(2):93–100

Yan Z, Du X, Fan X (2015) Numerical simulation of the evolutionary process of Danxia landforms. Phys Geogr 36(4):322–336
Yan L et al (2019) The spatial patterns of red beds and Danxia landforms: implication for the formation factors–China. Sci Reports 9(1):1961
Yanes CE, Briceño HO (1993) Chemical weathering and the formation of pseudo-karst topography in the Roraima Group, Gran Sabana, Venezuela. Chem Geol 107(3–4):341–343
Yang G et al (2012) Quartz sandstone peak forest landforms of Zhangjiajie Geopark, northwest Hunan Province, China: pattern, constraints and comparison. Environ Earth Sci 65:1877–1894
Young RW (1986) Tower karst in sandstone: bungle bungle massif, northwestern Australia. Z Geomorph 30:189–202
Young RW (2010) Bungle Bungle: tower karst in sandstone. In: Migoń P (ed) Geomorphological landscapes of the World. Springer, Dordrecht, pp 333–340
Young RW, Wray RAL, Young ARM (2009) Sandstone landforms. Cambridge Univ Press, Cambridge, p 314
Zerboni A, Perego A, Cremaschi M (2015) Geomorphological map of the Tadrart Acacus massif and the Erg Uan Kasa (Libyan central Sahara). J Maps 11(5):772–787

6 Karst and Other Limestone Landscapes

Abstract

Landforms associated with soluble rocks (e.g. limestone, dolomite, gypsum and halite) are generally referred to as karst, and they are among the most striking rocky landscapes in the world. They include a range of features such as gorges, underground drainage and caves, closed depressions (e.g. dolines), flutes (karren, lapiés), cones, pinnacles and towers. The nature of these rocks (solubility, porosity, permeability, jointing, bedding, etc.) contributes to the diversity of forms that is observed, but also significant is climate, past and present. Thus one can talk of cold climate karst, tropical karst, and dryland karst. Although solutional features predominate, in some environments one also has depositional forms, such as tufas, travertines, and spring mounds.

Keywords

Karst · Limestone · Solution · Dolines · Sinkholes · Poljes · Gorges

6.1 Introduction

Karst landscapes develop from a combination of high rock solubility and well-developed secondary permeability and are particularly associated with carbonate rocks such as limestones, marbles and dolomites. Not all limestones have well developed karst. Following the classification by the great great Serbian geomorphologist, Jovan Cvijić, Holokarst occurs where almost all the landforms, such as dolines, result from karstic activity. Here there is little or no surface runoff and there are few streams. The entire area is underlain by thick carbonate rocks. By contrast, merokarst is where karstic development is imperfect, as is the case on thin, impure, or chalky limestone where surface drainage and dry valleys are present in addition to some limited karstic features. Many areas are transitional between these two types (Ford 2006).

Goldscheider et al. (2020) provide a discussion of the distribution of carbonate rocks globally, regionally, and in terms of climate and topography. They suggest that 15.2% of the global ice-free continental surface is characterized by the presence of karstifiable carbonate rock. Karst has also developed in some other soluble rocks including gypsum/anhydrite and halite, as in the Cordillera de la Sal in the Atacama Desert in Chile (De Waele et al. 2009). Some karst-like phenomena, such as soil pipes in badlands, which are not the result of solution processes, are often called pseudo-karst, though this is a term that is not universally popular (Eberhard and Sharples 2013).

The term karst is derived from an area near Trieste on the Italy-Slovenia border. Karstic geomorphology is characterized by various types of closed depressions (e.g. dolines, cenotes, poljes, uvalas, etc.), deep gorges, underground drainage and caves, fluted rock outcrops (lapiés and karren) (see Fig. 6.1), residual hills (towers, cockpits) and the like. The effects of different limestones will depend on such factors as their bedding (see Fig. 6.2), their thickness, their density, and their jointing. Jennings (1985), Sweeting (1972) and Ford and Williams (2007) give full and expert reviews of karstic processes and forms. Veress (2020) provides a good general discussion of the factors affecting karst types at a global scale, while Britain is fortunate to have a series of regional studies of karstic landscapes (e.g. Ford 1977; Ford 1989; Smith and Draw 1975; Waltham 1974;Waltham et al. 1997). The landscapes of the classic dolomite terrain of northern Italy are described by Soldati (2010).

There is a huge literature on fluted solutional forms (see, for example, Ginés et al. 2009), but as many of them are too small to be examined on Google Earth, they are not for the most part described in this book, though exceptions, treated on account of their size, are the *Tsingy* of Madagascar and

A. Goudie, *Rocky Landscapes of the World with Google Earth*, https://doi.org/10.1007/978-3-031-77418-8_6

Fig. 6.1 Karren features, Lluch, Mallorca, Spain. See Mottershead (1996) for a discussion (Photo A.S. Goudie)

the Stone Forest of China. It needs to be remembered, however, that solution is not the only weathering process to affect limestones, and that they are also susceptible, as are other rock types to frost and salt weathering (Goudie 1999).

6.2 Gorges and Calanques

Many limestone areas are characterized by deep gorges, especially in areas that have undergone tectonic uplift. It is believed that they occur in karstic regions because river incision acts more effectively than slope processes. Some may be caused by collapse of the ground surface into underground voids, but this is unlikely except in the case of fairly small features. Most are caused by river erosion (fluviokarst). The Verdon Gorge (*Gorge du Verdon*) is located in south eastern France (Fig. 6.3). It is about 25 km long and up to 700 m deep. It was cut by the Verdon River (Cardinal et al. 2024). Other notable examples include the Tarn, Lot, Jonte and Dourbie gorges of France's Grands Causses, various gorges in the dolomites of Austria, in the Apuseni Mountains of Romania, Slovenský Raj in the Western Carpathians, south eastern Serbia, Cheddar Gorge in the Mendip Hills of England, the Samaria Gorge in Crete, and the Windjana and Geikie Gorges in the Kimberley district of Australia (see Sect. 6.7).

Also in France, to the east of Marseille and west of Cassis, are some gorges, which are also deep inlets of the sea (Fig. 6.4a, b). They are developed in Cretaceous (Urgonian) limestones. These are locally called *calanques* (Chardonnet 1948; Corbel 1956), though elsewhere they would be called rias. Some of the valleys, which were flooded to form calanques, might date back to the Messinian Salinity Crisis (5.96 and 5.32 mya). At that time time, the Mediterranean Sea became isolated from the Atlantic Ocean and its surface level dropped as much as 1500 m. As a result, rivers flowing into it deepened their valleys by hundreds of metres. The calanques have also been flooded by the post-glacial (Holocene or Flandrian) rise in sea level. Faults may have played a role in their formation as may have solution. Similar features are known from Malta and from Mallorca, where they are known as *calas* (Gómez-Pujol et al. 2013) (Fig. 6.5).

Fig. 6.2 Pancake Rocks, Punakaiki, New Zealand. (Photo the late M.M. Sweeting)

Fig. 6.3 The Verdon Gorge, France. 43°43′55.31″N, 6°23′19.85″E. ©Google Earth. Scale bar = 400 m

(a)

(b)

Fig. 6.4 **a** Calanque d'en Vau, France. 43°12′1.65″N, 5°30′14.61″E. ©Google Earth. Scale bar = 500 m. **b** Calanque near Marseille (Photo A.S. Goudie)

Fig. 6.5 Calas on the eastern coast of Mallorca. 39°20′58.54″N, 3°11′32.55″E. ©Google Earth. Scale bar = 400 m

6.3 Underground Drainage

Much drainage in karstic areas goes underground and flows through cave passages. Water flowing off impermeable rocks or from the snouts of glaciers, may plunge down potholes (sinkholes). Such valleys are often termed blind valleys. An example of this is Gaping Gill (Ghyll) in Yorkshire England (Fig. 6.6). Here the Fell Beck falls into a great cavern in the Great Scar Limestone, which was laid down as a sediment, almost 200 m thick, in a warm tropical sea, about 350 mya during the Lower Carboniferous period (Dinantian) when Europe and the British Isles lay about 10° south of the equator (Waltham 1974). The stream pours over a lip of rock above and crashes 98 m to the floor of the cavern before draining through to Ingleborough Cave. Gaping Gill has the highest unbroken waterfall in England and the largest underground chamber naturally open to the surface.

Ultimately, the process of karstification means that through time the surface drainage system will be destroyed and replaced by enclosed basins of all sizes and with more and more of the water being transmitted through enlarging cave systems. Cave systems transmit underground drainage and are also formed by it. Being underground, caves are not visible, except for some of their entrances, on Google Earth. An excellent review of caves is provided by Gillieson (2021). Underground karst is often termed endokarst, while surface karst features (karren, dolines, etc.) are often termed exokarst.

6.4 Dolines

Dolines, an archetypal landform class, are enclosed depressions found in karst landscapes. They are sub-circular in plan, tens to hundreds of metres in diameter, and can range from a few metres to about a kilometre in width. Some occur in hard limestones, but they can also occur on softer beds, such as the chalklands of southern England. Here some dolines have formed especially where sandy and acidic materials of Tertiary age overlie these soft Cretaceous limestones (Sperling et al. 1977). Their density reaches 157/km^2 on Puddletown Heath in Dorset, and one of the hollows, Culpepper's Dish has a mean diameter of 86 m. Their localization suggests that the dolines result from intense and localized solutional action under acidic conditions beneath heathland vegetation. They can still collapse into voids in the chalk, but many have probably

(a) (b)

Fig. 6.6 **a** Gaping Gill, Yorkshire, England. 54°9′0.13″, 2°22′56.81″W. ©Google Earth. Scale bar = 60 m. **b** Gaping Gill (Photo A.S. Goudie)

Fig. 6.7 The Xiaozhai Tiankeng doline in China. 30°45′1.99″N, 109°28′18.12″E. ©Google Earth. Scale bar = 300 m

developed by overlying materials being evacuated down solutionally enlarged joints due to an egg-timer effect. For the most part, howerver, the Chalk of England and France is characterized by merokarst.

While most dolines are typically a few metres to tens of metres in depth, some are hundreds of metres deep. One of the deepest examples is the collapse doline Xiaozhai Tinakeng (locally known as the 'Heavenly Pit') in China (Fig. 6.7). This is 511–662 m deep and c 600 m across. It formed over the Difeng cave, which in turn had been formed by a powerful underground river which still flows underneath the feature (Xuewen and Weihai 2006).

Enclosed depressions in karst can be formed by four main mechanisms: dissolution, collapse, suffosion and regional subsidence. In the first case, infiltrating rainwater is acidified in the atmosphere and further acidified in the soil and then flows into the limestone, widening the joints and forming a depression. Collapse dolines are formed mainly by mechanical processes as when the surface collapses into an underground cavern (Gabrovšek and Stepišnik 2011). Collapse refers to rapid downward movement of the ground, whereas subsidence refers to gradual movement sometimes without even ripping the surface. When unconsolidated deposits such as alluvium, glacial moraine, loess or sand mantle karstified rock, the sediments are sometimes evacuated downwards by suffusion through corrosionally enlarged pipes in the underlying karst, resulting in gradual or rapid subsidence of the surface. Figure 6.8 shows a useful classification of doline types based on Waltham et al. (2005) and previous authors.

Their sides range from gently sloping to vertical (see Fig. 6.9), and their overall form can vary from saucer shaped to conical or even cylindrical.

On the South Wales hills to the north of Ebbw Vale is an extensive area of dolines which occur in Carboniferous Limestone overlain by Milstone Grit. They are a type of collapse doline, sometimes called a *subjacent karst doline*. The Grit has the mechanical strength to form a relatively stable roof to caverns formed in the underlying limestone; these caverns are therefore able to grow to a considerable

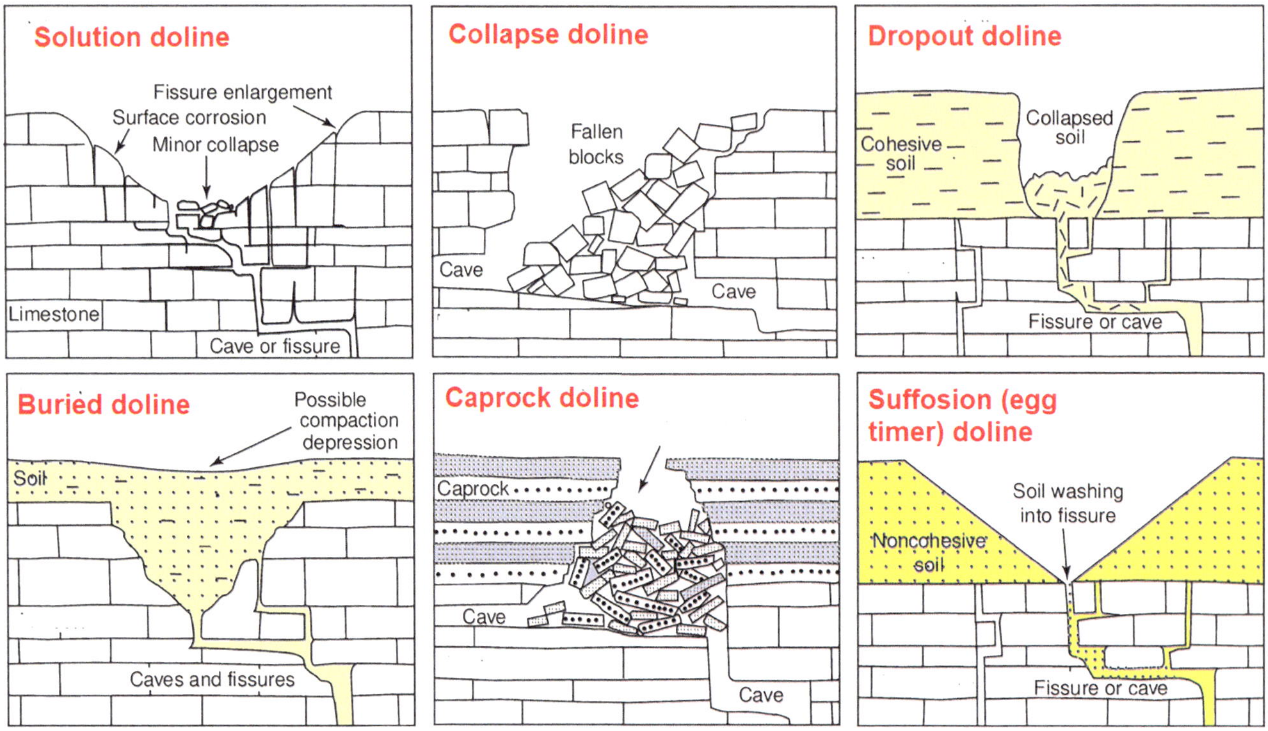

Fig. 6.8 Six types of doline (modified from Waltham et al. 2005, Fig. 2 and earlier sources)

Fig. 6.9 Ehjeri-Sdoose Doline, Jebel Akhdar, Libya (Photo A.S. Goudie). The dolines in this region are developed on Eocene to Miocene carbonates. Details are provided by Elshaafi et al. (2021)

size before collapse occurs. There are some thousands of dolines spread over about 60 km of outcrop, some with diameters in excess of 100 m. There may be as many a 50–75 features per ha (Thomas 1963). A striking example is Pwll Coch (Fig. 6.10).

Dolines are especially common in terrains underlain by carbonate rocks, but are also widespread on evaporites such as salt and gypsum (Gutierrez et al. 2008). Some are also found in siliceous rocks such as quartzite, as in parts of Brazil (Uagoda et al. 2011) and on the Roraima Plateau. The term sinkhole is sometimes used (especially in North America) to refer both to dolines and to depressions where streams sink underground, which in Europe are described by separate terms (including *ponor*, swallow hole, and stream-sink). Thus the terms doline and sinkhole are not strictly synonymous (Waltham et al. 2005). One name given to certain dolines is *cenote*. The type example occurs in the northern Yucatan Peninsula of Mexico, where the Mayan word *dzonot*, from which cenote is derived, means 'water cave'. The typical Yucatan cenote is a near-circular, water-filled shaft with vertical or overhanging walls extending up to 100 m downward from the ground surface (Moreno-Pérez et al. 2021). At Zacatón in Tamaulipas, Mexico, (22°59′34.14″N, 98°9′45.56″W) there is a 339 m deep cenote. Cenote-like flooded shafts, known as blue holes, occur in offshore reefs in the Caribbean (e.g. off Belize and the Bahamas) (Mylroie et al. 1995). These may perhaps be drowned cenotes (Fig. 6.11). The Great Blue Hole of Belize formed at time of low Pleistocene sea levels. It is located in the eastern lagoon of Lighthouse Reef, which is part of the Belize barrier and atoll reef system, and is 125 m deep and 320 m wide. It is surrounded almost completely by coalesced coral patch reefs (Gischler et al. 2017).

On the island of Gozo (Malta) is the large depression called Dwerja Bay (Fig. 6.12). The Maltese Islands, despite having an area of only 316 km^2, host a wide range of karst landforms (Calleja and Tonelli 2019). Among the most widespread are sinkholes, with over 50 having been recorded in various recent studies. The Dwejra area features a series of outstanding sinkholes that reach a few hundred metres in diameter and display a vertical displacement comparable in size to the largest sinkholes in the world. The Dwejra Bay sinkholes are currently flooded by the sea. A model of development of the Dwejra Depression through time was proposed by Gatt (2019). First, during the Pleistocene glacial period, sea level was 120 m below present, which exposed evaporites along the Eocene-Oligocene boundary to freshwater dissolution. The dissolution of gypsum/anhydrite formed an incipient cavern beneath the horizontally bedded Miocene Lower Coralline Limestone Formation. The subsidence of the limestone over the cavern created concentric faults, forming the structural low. Sea levels then rose in the Holocene and flooded this low-lying zone.

In Namibia, there is extensive karst development in the Etosha region, with many caves, solutional rills and cockpit karst (Goudie and Viles 2015a, b, p. 62). Close to Tsumeb are two unusual, deep, surface lakes, Otjikoto and Guinas. Otjikoto has a diameter of c 102 m and a depth in excess of 75 m (Fig. 6.13). It has formed in the carbonate rocks of the Neoproterozoic Otavi Group by the solution and collapse of surface rocks into a large sub-surface cavern. It is regarded as analogous in form and origin to the *cenotes* of Yucatan.

In Croatia, Crveno Jezero, the Red Lake, is a very impressive feature, and is amongst the deepest in the world (Garasic 2012). It has a vertical dimension of no less than 528 m, including the 280 m deep lake on its bottom (Fig. 6.14).

By contrast with these mega-dolines, in New Zealand's South Island, c 20 km inland from Timaru, is a doline field formed by subsidence into Oligocene limestones. It is partially smothered by loess. The dolines are small and numerous, with a mean diameter of only c 21.8 m (Jennings 1975) (Fig. 6.15).

On the Obruk Plateau in Anatolia, Türkiye, sinkholes, locally called *obruks*, have developed in Neogene lacustrine limestones (Fig. 6.16). Here there is an exceptional area hosting 67 hollows extending along a 50 km long and c 8 km wide rectilinear zone (Kuzucuoğlu 2019). Dogan and Yilmaz (2011) attributed the recent development of the *obruks*, not to dewatering associated with mining, which is a common cause of doline collapse, but to changing agricultural activity that has led to the opening of thousands of deep wells in recent years. Increased water pumping has exceeded the sustainable yield of the aquifer and has led to a sharp decline in the level of the water table. Waltham (2015) reported that well records showed that this has declined by about 24 m since 1980, and that most of the new *obruks* have occurred in the irrigated farmland. The implication is that the new collapses were induced by either collapse or suffusion (piping) or both.

About 15% of the continental USA has limestone, gypsum, or other soluble rocks at or near the surface, so karstic features are widespread. One major area is Florida, where depressions are numerous and cover large tracts of country (Beck and Sinclair 1986; Tihansky 1999) (Fig. 6.17). They are also hazardous, particularly when human actions, such as pumping of groundwater, modify hydrological conditions. Sinkholes can cause property damage and structural problems for buildings, roads, bridges, power transmission lines and pipelines and can also cause environmental problems such as degradation of groundwater quality in that open sinkholes can create pathways for transmitting contaminated surface water directly into the underlying groundwater aquifer. Many are water-filled, and have developed in Tertiary limestones. These formations are often covered or partially covered by sands and clays.

Fig. 6.10 The Pwll Coch doline, South Wales. ©Google Earth. 51°49′37.10″N, 3°12′43.25″W. Scale bar = 30 m

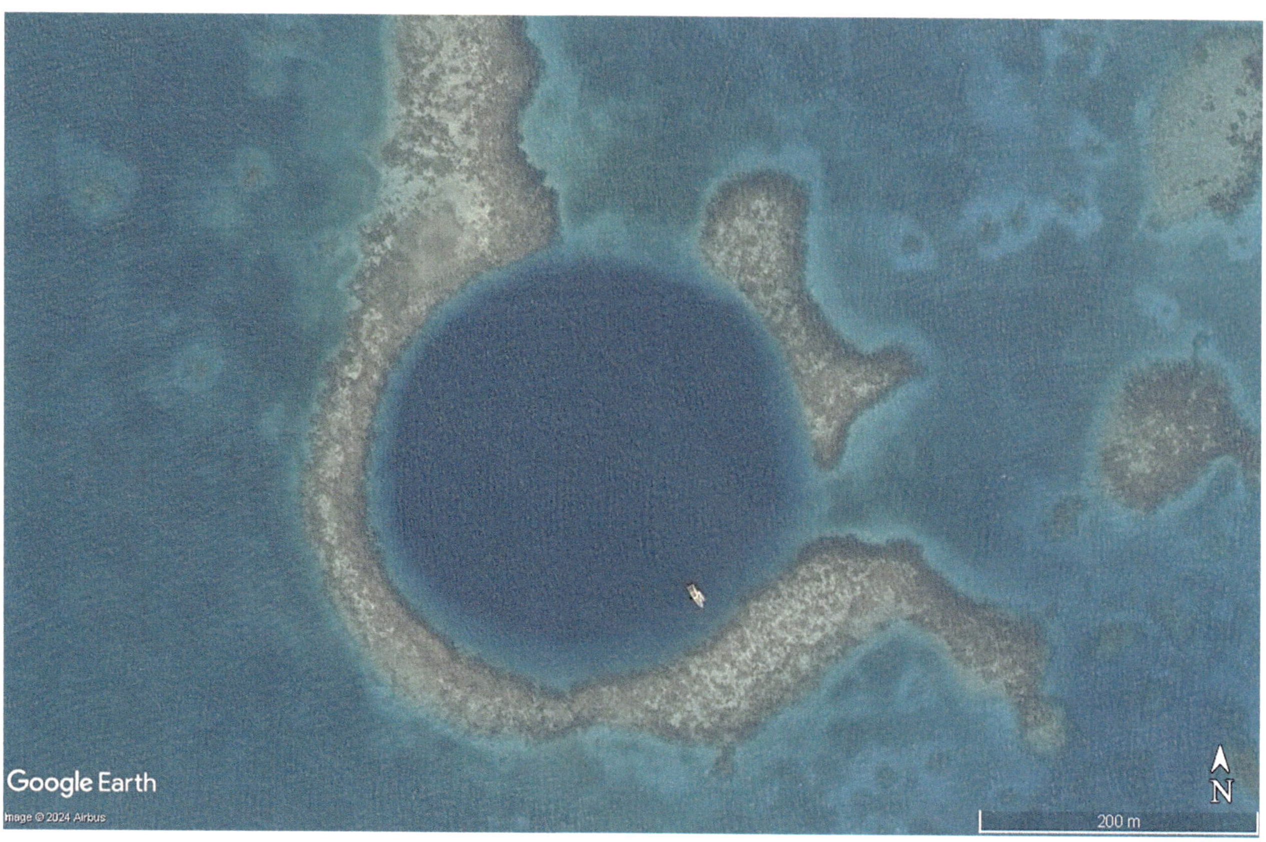

Fig. 6.11 The Great Blue Hole of Belize. 17°18′53.41″N, 87°32′7.76″W. ©Google Earth. Scale bar = 200 m

Fig. 6.12 Dwejra Bay, Gozo, Malta. 36°2′48.24″N, 14°11′28.13″E. ©Google Earth. Scale bar = 200 m. It is of similar width to the Belize Blue Hole

Some of the dolines may have formed through solution at times of low Pleistocene sea levels and have become lakes as sea level rose to its present level during the Holocene. Upchurch et al. (2019) have shown that there are multiple types of sinkhole in Florida: (1) *solution sinkholes* formed by dissolution of shallow limestone; (2) *suffosion sinkholes* where non-cohesive sediments creep down directly into limestone voids; (3) *rock-collapse sinkholes* formed by collapse of cave roofs; (4) *cenotes*, deep water-filled rock-collapse sinkholes; (5) *cover-collapse sinkholes* formed by the rapid failure of sand and clay over a limestone void; (6) *cover-subsidence sinkholes* formed by slow dissolution of the upper surface of a carbonate-rich sediment; and (7) *alluvial, or relict, sinkholes*. Not all the karstic depressions in Florida are sinkholes/dolines. Some of the larger examples may be poljes (Jensen 1987).

The Bimmah sinkhole in Oman is c 50 m by 70 m wide and approximately 20 m deep (Fig. 6.18). It is located in the coastal plain between Quriat and Fins. This is composed of gently north-dipping Eocene limestones (Seeb Formation) which are unconformably overlain by almost flat-lying lithified Pleistocene beach sands. The whole area is a classic karst with typical limestone weathering, riddled with caves and underground wadi systems that connect to the sea. The Bimmah sinkhole formed when Eocene limestones dissolved and collapsed above a deep cave system (Searle 2019a).

Dolines are not restricted to limestones and dolomites. They also occur on gypsum (Klimchouk et al. 1996) and anhydrite rocks. These relatively soluble materials are widespread and outcrop over c 7 million km^2 of the land surface, particularly in the USA's Permian basins (Land et al. 2003), Carboniferous areas of Nova Scotia, Canada (Fig. 6.19) (Baechler and Boehner 2014), Russia and the Mediterranean basin, including Spain (Calaforra and Pulido-Bosch 1999), Albania (Andreychouk et al. 2021) and Türkiye (Keskin and Yilmaz 2016).

The Turkish gypsum karst of the Sivas Basin is especially dramatic, displaying phenomena that range in size from small karren to large poljes (Fig. 6.20) (Doğan and Yeşilyurt 2019). It is one of the most outstanding gypsum karst terrains in the world, covering an area of 2140 km^2. The ENE–WSW trending Sivas gypsum karst terrain is 280 km long and 55 km wide. It is mainly developed on Oligocene gypsum deposits. As Doğan and Yeşilyurt (2019, p. 202) wrote:

(a)

(b)

Fig. 6.13 **a** Otjikoto, Namibia. 19°11′41.42″S, 17°33′0.09″E. ©Google Earth. Scale bar = 90 m. **b** Otjikoto, Namibia (Photo A.S. Goudie)

Fig. 6.14 Crveno Jezero doline, Croatia. 43°27′16.25″N, 17°12′1.76″E. ©Google Earth. Scale bar = 200 m

Fig. 6.15 The small Craigmore dolines, New Zealand. 44°26′16.10″S, 170°58′29.59″E. ©Google Earth. Scale bar = 300 m

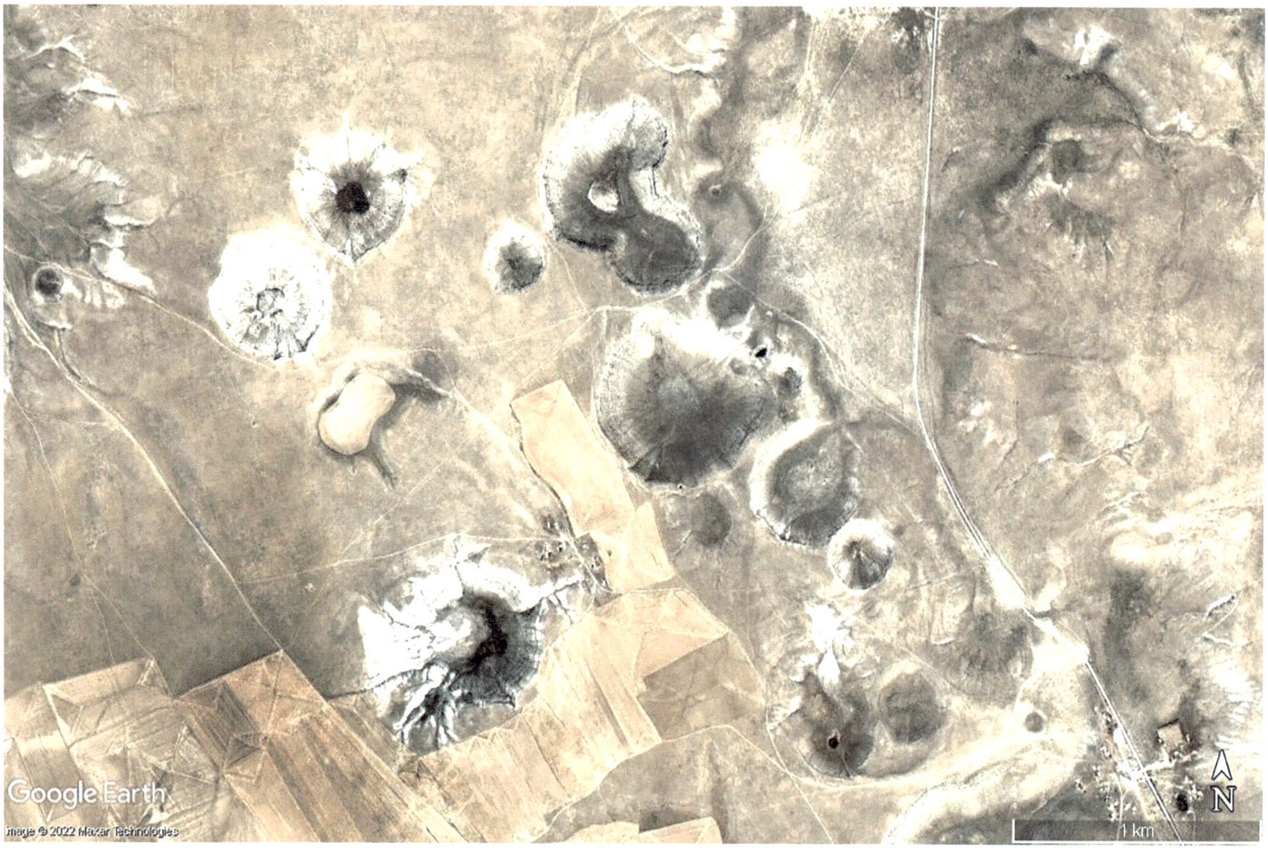

Fig. 6.16 Obruk Plateau dolines, Türkiye. 37°57′55.20″N, 33°23′8.60″E. ©Google Earth. Scale bar = 1 km

Fig. 6.17 Florida sinkholes, USA. 28°26′50.31″N, 81°31′57.01″W. ©Google Earth. Scale bar = 7 km

(a)

(b)

Fig. 6.18 **a** Bimmah sinkhole, Oman. 23°2′9.40″N, 59°4′19.08″E. ©Google Earth. Scale bar = 40 km. **b** Bimmah, Oman (Photo A.S. Goudie)

Fig. 6.19 Gypsum hollows at Cains Mountain, Nova Scotia, Canada. 45°57′41.42″N, 60°55′52.66″W. ©Google Earth. Scale bar = 400 m

> Most parts of the higher plateau constitute exceptional doline karst areas and display a truly outstanding karst landscape. There, the diameters of the dolines are between 10 and 500 m with depths down to 100 m in blind valleys. Some of the dolines have merged, acquiring an uvala form (compound dolines). The number of dolines per km^2 is high, reaching around 250–300, but most dolines are relatively shallow (<20 m) and small (<250 n diameter). Nearly all of them can be seen on satellite images. In the higher karst plateau, there is a polygonal network of low interfluve ridges that enclose shallow depressions with internal drainage through small sinks. The densely packed solution dolines in outcrops of bare gypsum form striking polygonal karst landscapes…

The polygonal depressions are sediment-filled and are often cultivated, as shown in Fig. 6.20b. The largest solutional feature, covering c 350 ha, is Lake Tödürge, but there are also a number of poljes (see Sect. 6.5) (Gökkaya and Gutiérrez 2022).

Evaporites, including halite (sodium chloride), are widespread both geographically (Klimchouk et al. 1996) and in the geological record (see Sect. 2.6 for discussion of salt structures). In areas where these deposits have accumulated, solution may cause sinkholes to develop. An example of this is provided by the Dead Sea basin, where the recent decline in level caused by water abstraction has promoted their formation (Yizhaq et al. 2017). Much subsidence has also occurred in the Ebro valley escarpment area of NE Spain (Migoń et al. 2023).

6.5 Poljes and Uvalas

Polje is a term used to refer to areas of flatland in the floors of large depressions. In the Dinaric karst of the Balkans they range in size from 2 to 400 km^2, but in that classic region most are less than 50 km^2. They tend to be aligned along structural trends (*structural poljes*). Indeed some may be related to faults. Others (*border poljes*) may have formed at the contact between impervious rocks and limestones, with acidic runoff from the former causing solution and lateral corrosion of the latter. A third type of polje—*baselevel poljes*—occur entirely in karst rock that has been incised by solution down to the level of the water-table (Williams 2004, Fig. 98). Water may drain into sinks called *ponors* or through a stream channel penetrating one of the walls. The flat floor of a polje may generally be covered by a layer of impermeable alluvium. Some poljes are wet, and contain lakes, as is the case with the gypsum poljes in the Sivas

(a)

(b)

Fig. 6.20 **a** Some gypsum karst features from the Sivas Basin of Türkiye. 39°53′55.40″N, 37°36′32.35″E. ©Google Earth. Scale bar = 2 km. In the north is an area pocked with polygonal solutional dolines, while in the south is Lake Tödürge, which is also believed to have a solutional origin. It possesses a submerged doline on its eastern shore. **b** Polygonal karst depressions in the Sivas Basin highlighted by areas of ploughing in 2018 and by dark tones. Exposed gypsum between the depressions is shown in a light tone. Black dots are trees. 39°54′56.91″N, 37°42′27.74″E. ©Google Earth. Scale bar = 200 m

Fig. 6.21 Cerknica Polje, Slovenia. 45°45′41.90″N, 14°23′22.29″E. ©Google Earth. Scale bar = 4 km

Fig. 6.22 Polje in cockpit karst area near Queen of Spains, Jamaica. 18°23′32.41″N, 77°45′49.49″W. ©Google Earth. Scale bar = 1 km.This is the largest polje in Jamaica

basin (see Sect. 6.4). About 130 examples have been found in the Dinaric karst of the Balkans, among them Cerknica Polje (Stepišnik et al. 2017) (Fig. 6.21). This polje is subject to winter flooding and is the largest polje in Slovenia.

Although the Balkans are the type site for poljes, notable examples are known from many other locations and climatic environments, including Jamaica (Lyew-Ayee et al. 2007) (Fig. 6.22), Spain and Morocco (De Waele and Melis 2009) (Fig. 6.23). They can also develop in gypsum, as in Türkiye (Gökkaya and Gutiérrez 2022).

The Dayet Ifrah polje in the Middle Atlas of Morocco is occupied by an endoreic lake with a surface area of 250 ha and a depth of 8 m. It is fed mainly by groundwater discharge, direct precipitation (snow and rain) and surface runoff during the wet season. The lake is situated on Lower Lias dolomitic limestone. To the south and north, the lake is bounded by ENE-WSW faults. The western side is characterised by the Tizi n'Tretten fault zone (Baadi et al. 2023).

As regards *uvalas* these are often defined as karst depressions which are formed due to the coalescence of several dolines. Cvijić (1901) proposed a direct genetic relationship, whereby sinkholes evolve by coalescence into uvalas and ultimately into *poljes*. However, Ćalić (2011) believes that this definition may well be wrong, and that no cyclic concept, involving a progression from doline to uvala and to polje, can be accepted. Instead, Ćalić concluded that uvalas are large (in km scale) karst closed depressions of irregular or elongated plan form resulting from accelerated corrosion along major tectonically broken zones (Fig. 6.24). Watson et al. (2019) also rejected the classic model after a study of closed depressions associated with evaporate collapses on the eastern shore of the Dead Sea.

6.6 Tropical Karst

Karst is well developed in tropical regions. Karstic dissolution may be the most intensive here on Earth, because of the large quantity of precipitation, high temperatures all year round, but mostly because of the great biological activity of the soil. If one were to over-generalise, one would argue that tropical karst is characterized by hills, whereas temperate karst is characterized by holes. This concept is discussed at length by Sweeting (1972, Chap. 15).

In the humid tropics roughly circular dolines may be replaced by irregular star-shaped hollows with rounded hills (cones). They have a range of different names, including 'haystack hills', 'Pepino Hills', 'tit hills', and *kegelkarst*. Those in Puerto Rico are called *mogotes* (Day 1978) (Fig. 6.25), while In Jamaica these features are known as *cockpits*. Similar features are also known from such locations as Belize, the Philippines and Indonesia (Kusumayudha et al. 2015). The Chocolate Hills area in Bohol in the Philippines has about 2000 hillocks. Their mean heights are from 40 to 120 m and their diameters are about 65 to 80 m (Salomon 2011) (Fig. 6.26). Bohol is a UNESCO Global Geopark.

Those in Jamaica are often 50 m in depth and may be as much as 1 km in diameter. The polygonal depressions in the Cockpit Country in the northwest of the country cover about 600 km^2. They seem to have developed the intersection between master joint sets (Lyew-Ayee 2010), and most occur in the Eocene White Limestone Formation. They are partly moulded by fluvial processes associated with high intensity rainfall, but some have been toughened by case hardening (Fig. 6.27). Indeed, Lyew-Ayee (2004) found that fresh Montpelier limestone had Schmidt Hammer R values that averaged as little as 14, whereas the case hardened zones had R values as high as 38. He suggested that this enabled towers to form in what was otherwise a very weak material. Day (2004, p. 233) has remarked that 'The term "cone karst" is in some ways inappropriate, since the residual hills are rarely conical or isolated, rather being linked at their bases as irregular ridge remnants, and it is the enclosed depressions rather than the hills, which are the focus of geomorphic activity'. They are clearly rather different in morphology from The Chocolate Hills in the Philippines.

Cone karst occurs in some parts of China (Waltham 2008) as in the Luoping area of Yunnan. This has been likened to egg box topography and is locally called *fencong*. It is possible that as incision and weathering continue, the hills may become wider and deeper and metamorphose into towers (*fenglin*) (Fig. 6.28), which may themselves become degraded through time into modest residual hills and plains.

Elsewhere, the hollows may show more linear forms, and this produces what is known as labyrinth karst (Brook and Ford 1978). Examples are known, for instance, from Indonesia (Fig. 6.29) (Haryono and Day 2004), and joint control appears to be paramount in their development.

In some lower latitudes, the karst may be composed of a series of imposing towers or inselbergs, 30–200 m high and with near vertical walls. Tower karst (also known as *turmkarst*) has been recognized in massive limestones in, *inter alia*, Sarawak, Malaysia, Cuba, Puerto Rico, and in the Guilin area of China. In Vietnam, especially in Ha Long Bay, tower karst has been flooded by sea level rise (Waltham 2005) (Fig. 6.30a). Here there are more than 3000 towers and isles (Cerrano et al. 2006). Similarly impressive drowned cone karst and dolines occurs on the Island of Palau in the Caroline chain of the western Pacific (Waltham 2019) (Fig. 6.30b).

The tower karst of China is especially striking, particularly near Guilin (Daoxian 1987; Sweeting 1995) (Fig. 6.31). Many of the towers (*fenglin* karst) sit in broad alluvial plains that contain flat-floors and swampy depressions, while others rise up above bedrock plains formed in other lithologies. They have developed in Devonian and Cretaceous limestones and have been associated with long continued tectonic uplift (Waltham 2010). Indeed, it is the combination of great

Fig. 6.23 Dayet Ifrah polje, Morocco. 33°33′28.31″N, 7°33′54.76″S. ©Google Earth. Scale bar = 1 km

Fig. 6.24 Lomska Duliba uvala, Croatia. 44°46′30.71″N, 15°1′21.55″E. ©Google Earth. Scale bar = 600 m

Fig. 6.25 *Mogotes* at Manati, Puerto Rico. 18°24′10.97″N, 66°28′2.74″W. ©Google Earth. Scale bar = 1 km

Fig. 6.26 Cone karst in the Chocolate Hills, Philippines. 9°48′45.54″N, 124°8′55.83″E. ©Google Earth. Scale bar = 500 m

(a)

(b)

Fig. 6.27 **a** Cockpit Country, Jamaica. 18°20′24.56″N, 77°18′57.21″W. ©Google Earth. Scale bar = 1 km. **b** Cockpit Country, Jamaica (Photo the late M.M. Sweeting)

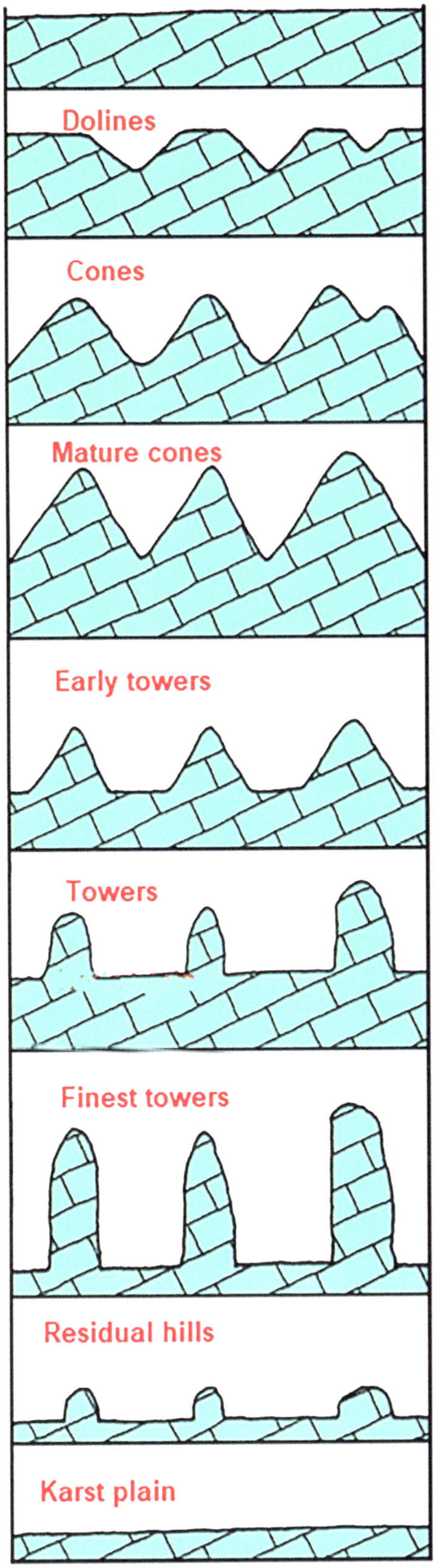

Fig. 6.28 The evolution of cone (*fengcong*) and tower karst (*fenglin*). Modified after Waltham (2008, Fig. 19)

expanses of thick limestones (exceeding 3000 m) and slow epeirogenic uplift during the Cenozoic Era that have made the kast of South China so exceptional.

Another fine display of Chinese tower karst is that in south-central Guizhou (Fig. 6.32). This is an area with strong neotectonic uplift and this has contributed to the development of the striking landforms (He 1987).

Another type of Chinese karst, is the stone forest. This consists of pointed pillars between 5 and 50 m in height. These are formed in fissured Lower Permian carbonates (see Song and Liang 2009) (Fig. 6.33). The size of the pillars in the Naigu area of Yunnan is mainly the consequence of faults and fissures that vertically criss-cross the beds, and the effectiveness of subsoil dissolution of the rock along them (Knez and Slabe 2011).

In Madagascar, there is a distinctive area of karst called *tsingy*. This appears to be very similar to the stone forests of China. Other examples are known from Cameroon, Congo, coastal Kenya, Brazil, Thailand, Sarawak, Papua New Guinea, and tropical Australia. Madagascar is rich in limestones and the tsingy are 'among the most extraordinary karstic landscapes of the planet' (Salomon 2009, p. 411). Tsingy karsts cover c 800 km^2 and those of Bemaraha are a World Heritage Site. The factors responsible for their development are very pure, low porosity limestones, a very strong fracture development, a relatively slight dip (<5°), and heavy and concentrated rainfall (Fig. 6.34).

In the humid tropics, karst is by no means confined to limestones. Karstic forms occur on the quartzite tepuis of Roraima, South America (see Sect. 5.2). They have also been found in peridotite, an olivine-rich, ultramafic igneous rock, in New Caledonia (Jeanpert et al. 2016). There one finds large fields of predominantly small dolines (Fig. 6.35), associated with laterite weathering profiles on top of the peridotite. Wirthmann (2000) also draws attention to what he calls semi-poljes. The peridotite, when at the surface, appears to be very unstable under tropical conditions. Karst developed in laterite is also quite extensive in the humid regions of northern Australia (Grimes and Spate 2008). Karren and caves have also been reported from granite and gneiss in Sri Lanka (Osborne et al. 2013) and from sandstones and quartzites in regions including Brazil (see Sect. 5.2.8) (Willems et al. 2008), the central Sahara (see Sect. 5.2.7) and northern Australia (Grimes 2012).

6.7 Windjana Gorge—Karst in Tropical Northwest Australia

The Kimberley area in northwestern Australia has some very notable karst that was first described in detail by Jennings and Sweeting (1963). It has pinnacle karst

Fig. 6.29 Labyrinth karst, Indonesia. 8°5′50.17″S, 110°29′30.72″E. ©Google Earth. Scale bar = 600 m

similar to that of China and Madagascar, gorges, and large expanses of tufa, but is most notable for the fact that the Napier Range is an exhumed Devonian reef in an area of intense summer rainfall.

In the Devonian, about 350 mya, a U-shaped barrier reef surrounded the Kimberleys. The reef complex, now remaining as the Napier Range, the Oscar Range and the Ningbing Range near Kununurra, skirted around the Kimberley on the northern margin of the Canning Basin. Round about c 20 mya, the whole area began to be uplifted and eroded, exposing the reef which had been buried under later sediments. The limestone forming the reef was more resistant to erosion than these overlying rocks, so that the ancient reef landscape was exhumed from beneath them.

The ancient reef is composed primarily of three types of hard limestone: a reef facies (the Windjana Limestone), a back-reef facies (the Pillara Limestone), and a fore- and inter-reef facies (the Napier Formation).

Slope forms, from near-vertical cliffs to gentle convexo-concave profiles, can be related to these different facies and their geotechnical properties (Allison and Goudie 1990; Allison et al. 1993). The Napier Range is a sinuous NW–SE trending ridge (Goudie et al. 1990) (Fig. 6.36). It extends over a distance of approximately 120 km, is up to 5 km wide, and is drained by the series of rivers which flow to the Indian Ocean. The maximum relief is of the order of 80–100 m. The preservation of a palpable reef form of such an age is remarkable. In 'the classic face' in the Windjana Gorge, cut by the Lennard River, one can see the various facies well displayed (see Playford 1980; Playford et al. 2009) (Fig. 6.37).

For comparison one can look at a current day barrier reef – the Great Barrier Reef of Queensland, Australia. It is broadly of the same great length as the Devonian Reef of Kimberleys. It is the world's largest coral reef system and is composed of over 2900 individual reefs and 900 islands stretching for over 2300 km (Hopley et al. 2007) (Fig. 6.38).

6.8 Cold Karst

In cold regions, karstic forms can be different from those in other climatic zones because of the past or present roles of ice. Many of them show the impact of glaciation (glaciokarst) and of frost weathering and other periglacial processes (Veress et al. 2019). The glacial impact is in some respects like that for other rock types, with cirques, nunataks, moraines, and the like. However, spectacular caves, canyons, labyrinths and sinkholes can form in cold regions,

(a)
(b)

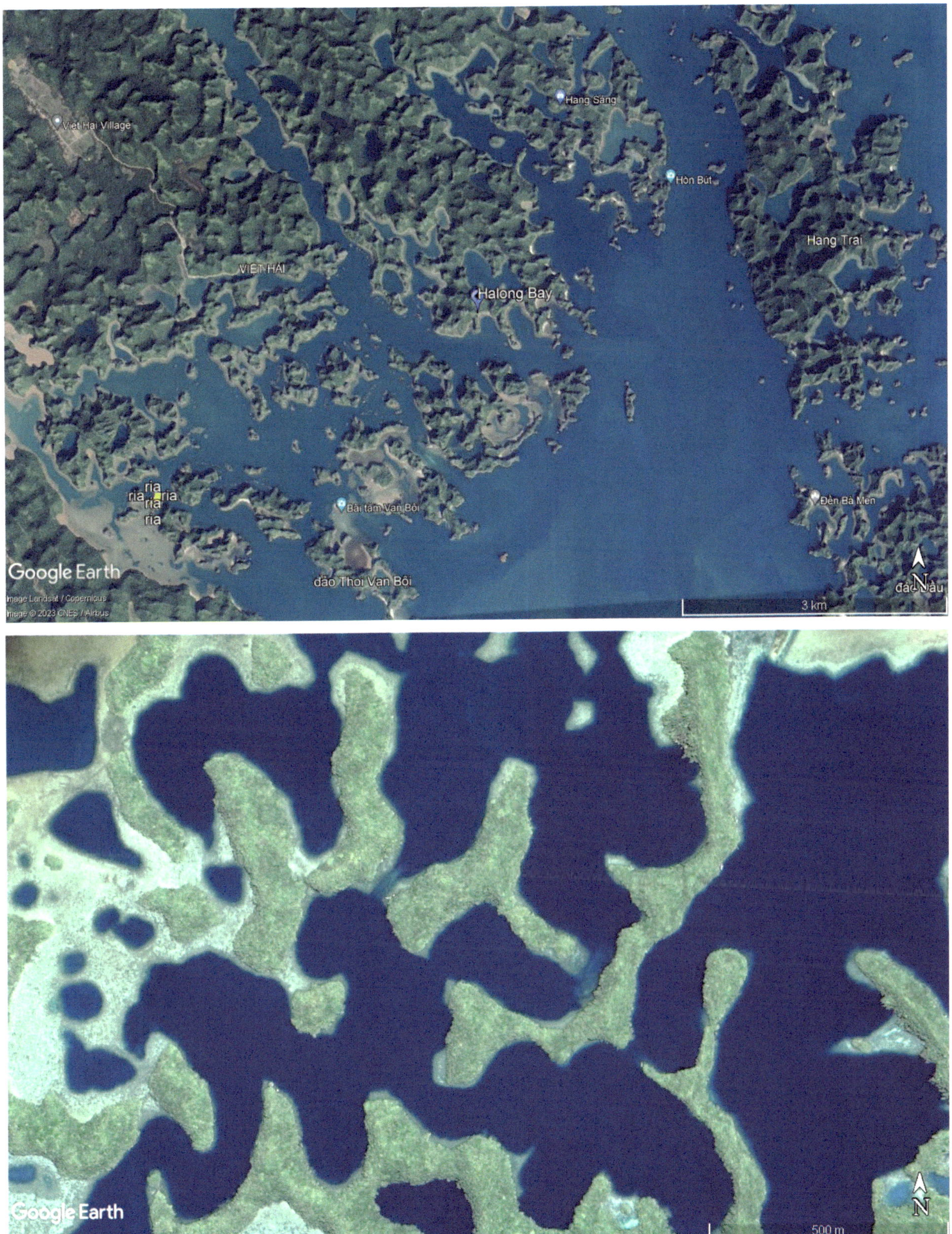

Fig. 6.30 **a** Drowned karst at Ha Long Bay, Vietnam. 20°46′45.35″N, 107°5′20.64″E. ©Google Earth. Scale bar = 3 km. **b** Drowned karst on Palau with dolines (left) and irregular fencong. 7°19′46.64″N, 134°29′24.56″E. ©Google Earth. Scale bar = 500 m

(a)

(b)

Fig. 6.31 **a** Guilin tower karst, northern Guangxi, China. 25°10′55.30″N, 110°14′59.10″E. ©Google Earth. Scale bar = 3 km. **b** Guilin tower karst (Photo the late M.M. Sweeting)

Fig. 6.32 Tower karst in south-central Guizhou, China. 25°1′0.78″N, 110°25′25.47″E. ©Google Earth. Scale bar = 4 km

as Ford (2010, 2016) has demonstrated for the Nahanni Karst in the Northwest Territories of Canada.

An example of cold karst is provided by the Tsanfleuron area in the Swiss Alps (Fig. 6.39) where the upper zone is still occupied by a 4 km^2 glacier, a middle zone shows the impact of expanded glaciation in the Little Ice Age, and the whole area shows the impact of Pleistocene glaciation (Veress et al. 2009). Since about 1860, the glacier has been retreating from a moraine ridge which is now located about 1.6 km from the ice front. Away from the presently glaciated area there are extensive karren fields, and expanses of shattered bedrock, blockfields and screes (Reynard et al. 2003). Large 1–8 m-high bedrock steps called *Schichttreppen* karst (stepped-pavement karst) dominate the area. Some of these steps are more than several hundred metres wide. The surface of the Cretaceous limestones also shows the impact of sub-glacial meltwater flows in the form of an intricate pattern of channels (Sharp et al. 1989). Active swallow holes occur where abundant supra- and sub-glacial meltwater enters the karst system near the glacier front (Steinemann et al. 2020).

Past glacial erosion often scoured bedrock surfaces, leading to the formation of roches moutonnées and limestone pavements that have been moulded by post-glacial weathering to produce various types of karren and solutionally-widened joint systems. Examples of this are the Hutton Roof Crags and the Goredale/Malham area of northwest England. The pavement surfaces (Fig. 6.40a) may sometimes be partially covered in glacial debris and post-glacial solution may have lead to the development of perched erratic boulders, as at Norber (Parry 2020) (Fig. 6.40b). The erratics were emplaced about 18,000 years ago and since then the limestone pavement may have been lowered by as much as 22–45 cm (Wilson et al. 2012).

A particularly spectacular example of a landscape moulded by past glaciation is the Burren area of County Clare (a UNESCO Global Geopark) and the Aran Islands of western Ireland. The former (Fig. 6.41) is composed primarily of Lower Carboniferous limestone, deposited in shallow equatorial seas 354–326 mya and uplifted in several phases over the last 250 my. It was glaciated at least twice during the Pleistocene. Here one sees not only the stripped surfaces produced by Pleistocene glaciation, but also streamlined surfaces created by glacial abrasion, and glacially excavated depressions. Limestone pavements are the most characteristic and widespread large-scale karstic landform of the Burren, though there are also in excess of 1500 dolines with an area greater than 100 m^2. Also of note is the Carran Polje, which is the largest closed karstic depression in the British Isles (Hennessy et al. 2017).

Fig. 6.33 **a** Stone forest, China. 24°48′41.80″N, 103°19′28.98″E. ©Google Earth. Scale bar = 90 m. **b** Stone forest in China (Photo the late M.M Sweeting)

Fig. 6.34 Tsingy de Namoroka, Madagascar. 16°26′42.71″S, 45°19′39.68″E ©Google Earth. Scale bar = 70 m

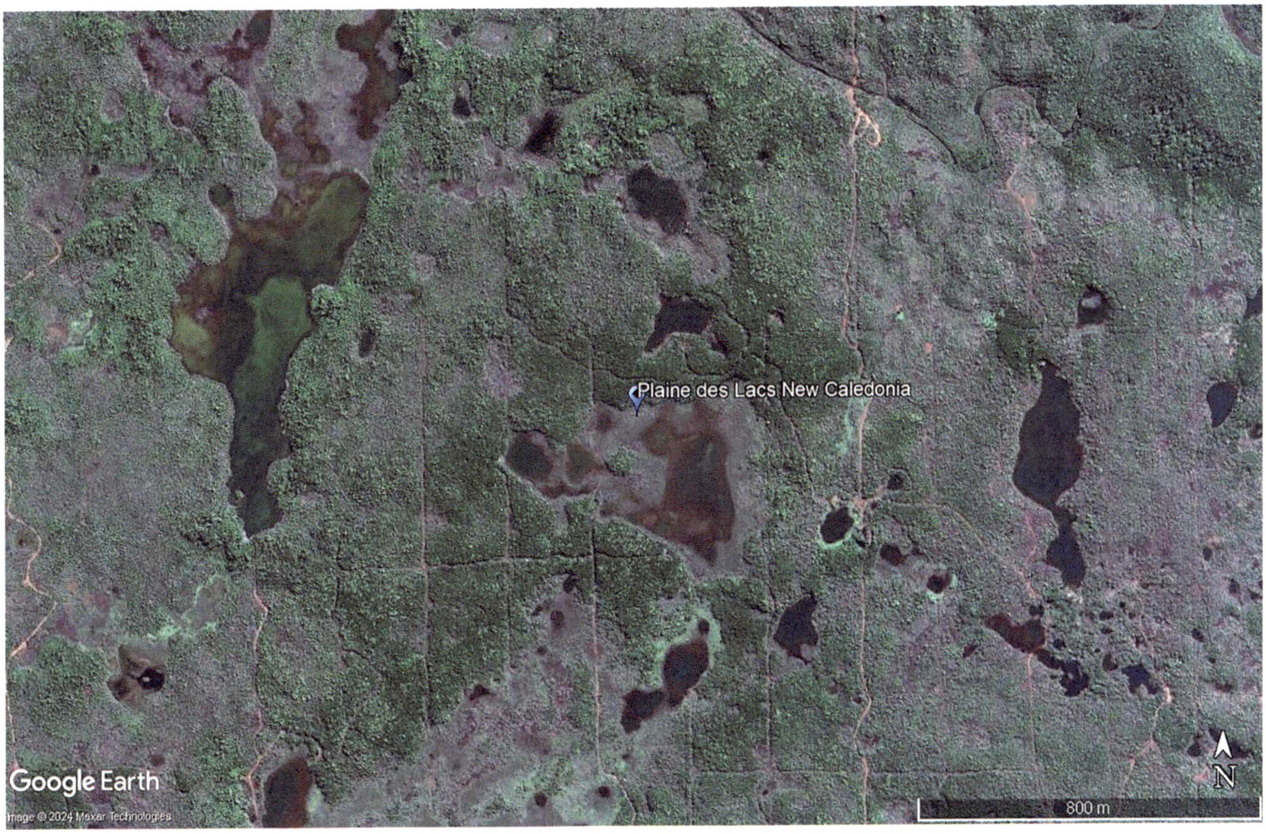

Fig. 6.35 Peridotite dolines, Plaine des Lacs, New Caledonia. 22°15′27.75″S, 166°56′33.17″E. ©Google Earth. Scale bar = 800 m

(a)
(b)

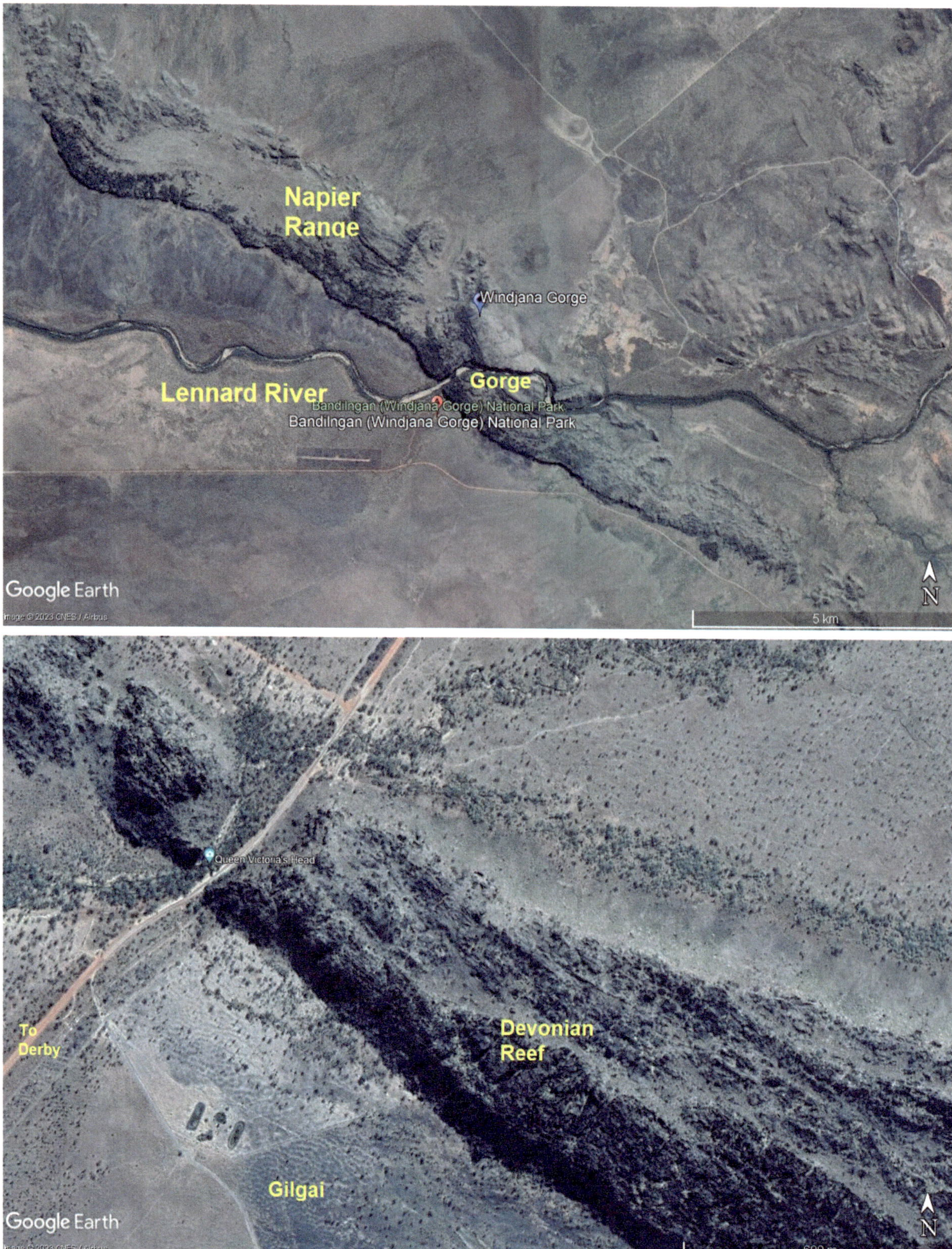

Fig. 6.36 **a** Windjana Gorge, Western Australia. 17°23′52.64″S, 124°57′1.47″E. ©Google Earth. Scale bar = 5 km. **b** Napier Range, Western Australia. 17°18′27.02″S, 124°47′34.79″E. ©Google Earth. Scale bar = 600 m. **c** The abrupt face of the reef at the Morown cliffs, Oscar Range, Kimberley (Photo A.S. Goudie)

(c)

Fig. 6.36 (continued)

Holocene landscape evolution in the Burren appears to be appreciably biokarstic in origin, rather than purely karstic, with biological processes playing a key role in landscape development in the last *c.* 13,000 years and with the formation of many micro-forms (karren) (de la Rosa 2016).

The three Aran Islands lie approximately 10–14 km off the southwest coast of Galway (Fig. 6.42). The rock comprising the islands is pure-bedded limestone of Lower Carboniferous age, and geologically the Aran Islands are an extension of the Burren. Boulder erratics of granite carried from the mainland at Connemara are dotted across the glacial pavements of the islands, and signs of glacial stripping and moulding are widespread.

Under periglacial conditions, surface runoff may have been promoted by snow melt and by glacial meltwater, leading to gorge incision and the formation of dry valleys. Surface runoff would also have been promoted by the normally permeable limestones being rendered impermeable by the presence of permafrost. In addition, bedrock would have been fractured by frost weathering and solifluxion would have been active (Whiteman and Haggard 2018). Dry valleys are widespread in the merokarst of southern England, especially on the Cretaceous Chalk. One example of this is The Manger at Uffington (Fig. 6.43). On its steep sides are some near parallel stubby chutes that may have been formed by snow avalanches. In addition. Great spreads of chalk debris produced by frost activity and by solifluction occur at the foot of the escarpment as a series of large fans (Paterson 1977; Goudie and Gardner 1985, p. 119–120).

6.9 Dayas and Dryland Karst

Under conditions of aridity, karst either may be ill-developed and retarded or be the relict of past more humid eras. This is the case in the great Nullarbor Plain of southern Australia, where there is currently no surface drainage, but caves, dayas, dolines, and pocket valleys that may date back to the Miocene and Pliocene (Lipar and Ferk 2015; Woodhead et al. 2019).

Dayas, often not more than a few metres deep, but measuring from tens of metres up to several km in diameter, are shallow depressions which are sometimes found

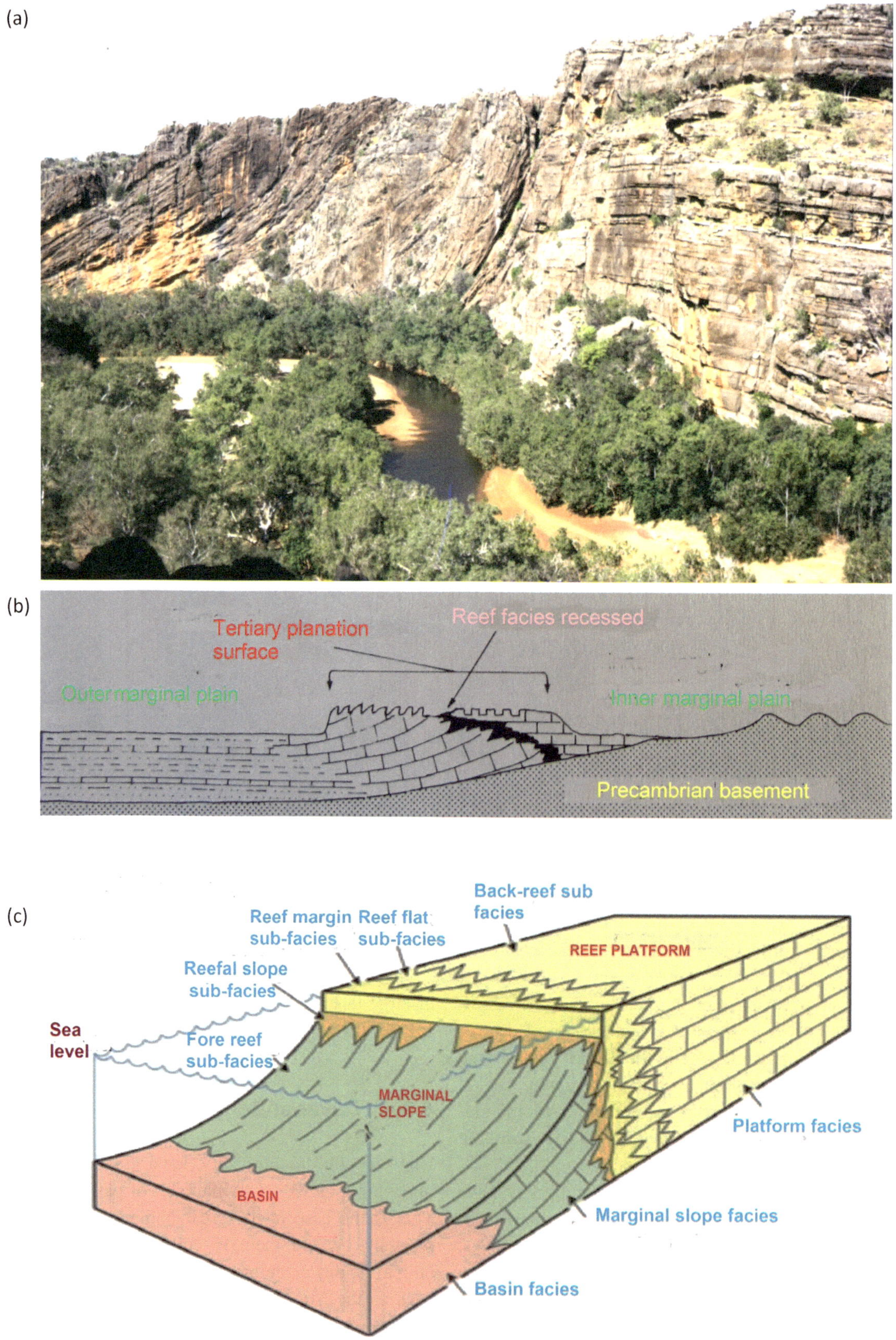

Fig. 6.37 **a** The 'classic face', Windjana Gorge, shows the structure of the former barrier reef in cross-section (Photo A.S. Goudie). **b** The present relief at Windjana Gorge (modified by author from Jennings 1985, Fig. 4). **c** Playford's Model of Windjana Gorge's 'classic face' (modified by author from Playford 1980)

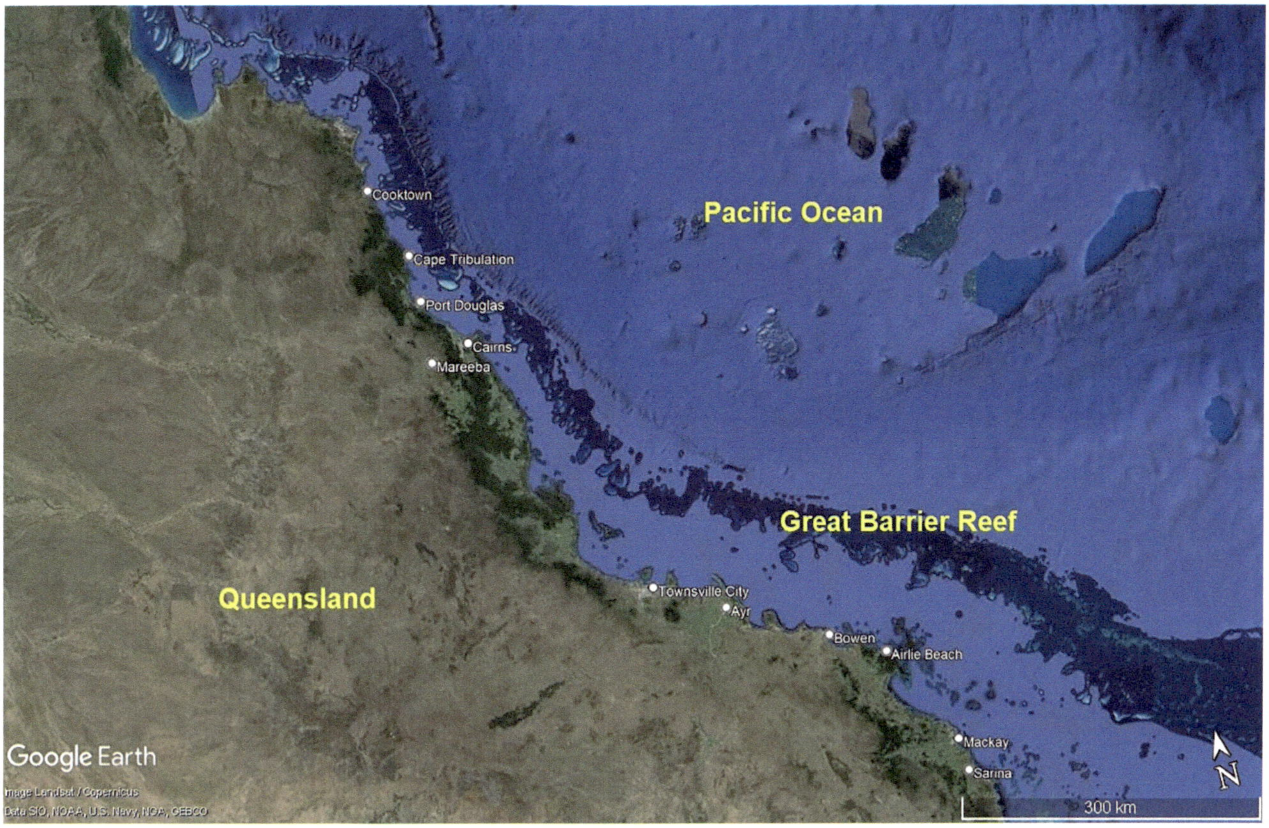

Fig. 6.38 The Great Barrier Reef, Australia. 18°31′47.04″S, 147°1′2.36″E. ©Google Earth. Scale bar = 300 km

scattered over limestone surfaces in arid regions (Mitchell and Willimott 1974). Large areas of dayas have been found in many drylands, including the High Plains of Texas, the provinces of La Pampa and Buenos Aires in Argentina, the Weissrand Plateau of Namibia, South Africa, northern Kenya, the Nullarbor Plain of Australia, Senegal, Mauritania, Western Sahara, Morocco, Algeria, Libya, Egypt, Ethiopia, Somalia, Bahrain, Kuwait, Iraq, Qatar and Saudi Arabia. Their presence on calcareous rocks, including calcretes, suggests very strongly that karstic processes may be involved in their development and that they may therefore be a type of doline. Because they are normally such shallow features, it is often difficult to locate or identify them on the ground, but in recent years the availability of remote sensing imagery makes it possible to gain a fuller appreciation of their distribution and forms (Goudie 2010).

Images of the Weissrand plateau area of southeastern Namibia (Fig. 6.44) show that there is a large area of broadly parallel drainage lines along which there is a multitude of small closed depressions (Goudie and Viles 2015a, b). This area runs approximately 150 km from northwest to southeast and 50 km from south west to north east. The depressions are numerous, with approximately 2 per km^2. They tend to occur in lines running from northwest to southeast, with around 1.3–1.4 km between depressions. The depressions themselves have lengths that range between about 100 and 1200 m, with the mean being around 460 m.

They have developed on calcretes which cap Ecca (Permian) sedimentary rocks (shales, mudstones, etc.) and later Kalahari sediments, which include not only calcretes but also pan deposits, and aeolian sands of various ages.

There is a striking near correspondence between the alignment of the Weissrand drainage and the alignment of the linear dunes and inter-dunal depressions of the Kalahari. The drainage lines have a spacing and parallel alignment which implies (i) that they may have been superimposed onto the calcrete by a formerly more extensive dune cover and (ii) that the closed depressions may have developed along the drainage lines because of localized solution of the underlying calcrete, assisted perhaps by other pan-forming processes such as deflation and animal activity. They may have evolved through the following four stages (Goudie 2010). In Stage 1 there is an extensive calcrete surface over which there is a mantle of linear dunes. Some very modest depressions develop along the interdune swales, either as a result of deflation or solution, or a combination of the two, but most of the calcrete is not directly exposed to the atmosphere. However, in Stage 2 the dunes become further stripped off, either because of a reduction in sand supply,

(a)

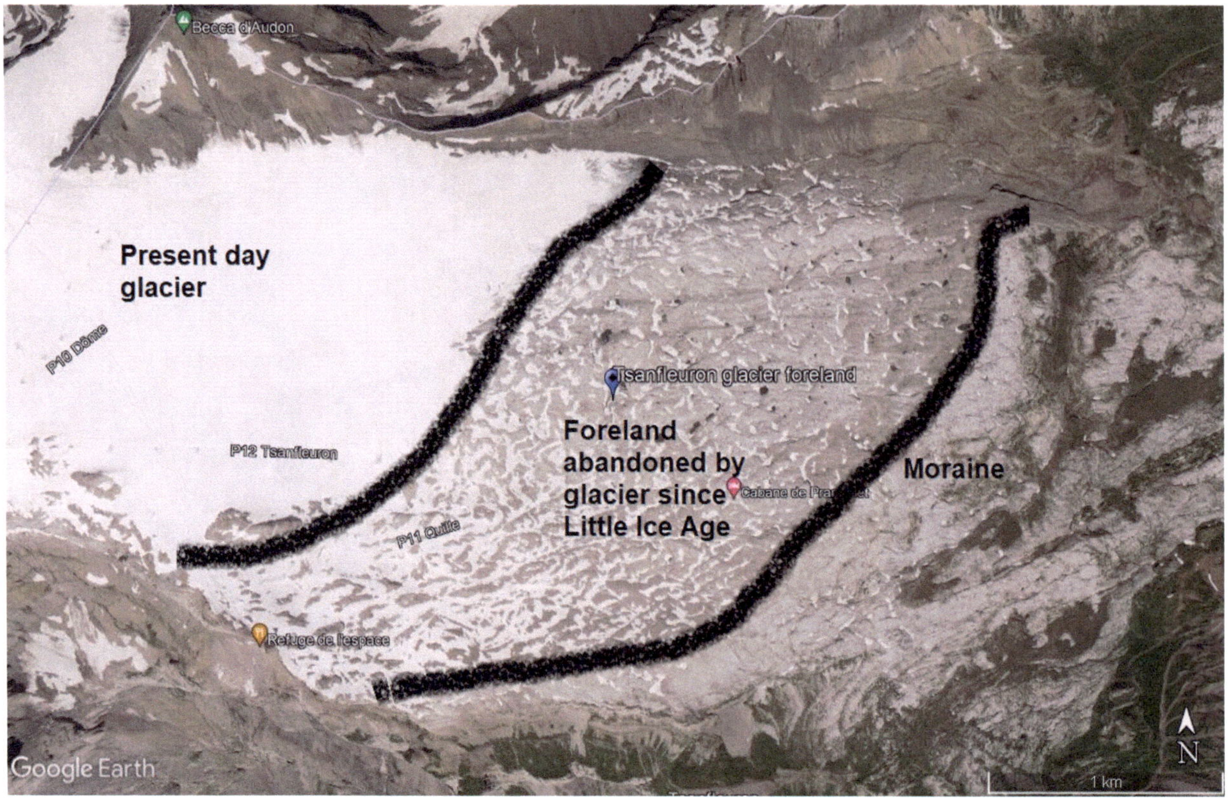

(b)

Fig. 6.39 **a** Tsanfleuron Glacier Foreland, Bernese Alps, Switzerland. 46°19′3.15″N, 7°14′29.35″E. ©Google Earth. Scale bar = 1 km. On the left is the present glacier, in the middle is the foreland, and on the right is a moraine ridge. **b** Tsanfleuron, Bernese Alps, Switzerland showing the extensive development of shattered bedrock and screes. 46°18′3.20″N, 7°14′40.85″E. ©Google Earth. Scale bar = 600 m

(a)

(b)

Fig. 6.40 **a** Limestone pavement with karren features (clints and grykes), Goredale, Yorkshire, England (Photo A.S. Goudie). **b** Norber glacial erratic (Silurian sandstone) on Carbonifeous limestone pedestal, England (Photo A.S. Goudie)

Fig. 6.41 Burren, Ireland, stripped limestone surface. 53°8′21.57″N, 9°2′42.35W. ©Google Earth. Scale bar = 200 m

or because of a shift in wind direction. This exposes the calcrete surface to the atmosphere, so that the depressions that had developed in Stage 1 form an aligned pattern that is etched into it. In Stage 3 the depressions are developed further as a result of solution, and themselves start to become foci for shallow drainage lines, which means that a more dendritic pattern develops. In Stage 4, when the calcrete caprock becomes breached, the underlying Ecca shales are subjected to such processes as salt weathering and deflation, and eventually large, characteristically shaped pans evolve, some with lunettes on their lee (south east) sides. A centripetal drainage pattern develops, focused on the pan.

Dayas have formed on silicified Neogene limestones in the otherwise featureless plateau of the Hamada Du Guir in the far west of Algeria (Castellani and Dragoni 1986; Thiry and Ben Brahim 1997) (Fig. 6.45) and in neighbouring parts of south east Morocco (e.g. at 31°33′29.92″N, 3°41′44.95″W).

In addition to dayas, dryland areas have some distinctive features, including deposition of secondary carbonates (tufas, etc.), groundwater sapping, and wind erosion forms (Fig. 6.46).

The Kharga depression of the Libyan Desert in Western Egypt, is a closed basin c 200 km across and 70 km long cut into very gently-dipping sedimentary rocks. Particularly on its eastern side, it is overlooked by steep cliffs of Late Cretaceous-Early Eocene limestones, and into these are cut saw-shaped valleys. At the heads of some of these valleys, as at Refuf, are ancient deposits of calcareous tufas, which have been precipitated by springs (Mousa et al. 2024). The valleys are stubby, and there is minimal stream development evident on the plateau surfaces (Fig. 6.47). On the basis of these characteristics, and using satellite images, Luo et al. (1997) postulated that the drainage net was created by spring sapping. This was a view that was confirmed, not only for Kharga, but also for Farafra and Qattara, by Abotalib et al. (2016). Other river networks that have possibly been created by groundwater sapping are known from the Gilf Kebir in Egypt, the Colorado Plateau of the USA, the Kalahari, Inner Mongolia and Chile. That said, Lamb et al. (2006) have cautioned that some of these features may have been formed by flash floods.

The geology of Kharga favours spring sapping, for the basin floor is composed mainly of impermeable Cretaceous shales, while the plateau surface is composed mainly of permeable early Eocene limestones. It is likely that sapping, slope slumping, scalloping, and tufa deposition occurred when wetter conditions existed. At the moment

(a)

(b)

Fig. 6.42 **a** Aran Islands, Ireland, showing stripped limestone surface. 53°8′28.10″N, 9°47′35.42″W. ©Google Earth. Scale bar = 700 m. **b** Glacially-moulded limestone, Aran Islands (Photo A.S. Goudie)

(a)

(b)

Fig. 6.43 **a** White Horse dry valley, Uffington, Oxfordshire, England. 51°34′46.08″N, 1°34′14.04″W. ©Google Earth. Scale bar = 300 m. **b** Uffington avalanche chutes (left) on the side of The Manger dry valley (Photo A.S. Goudie)

Fig. 6.44 Dayas, Weissrand Plateau, Namibia. 25°45′32.56″S, 18°44′7.90″E. ©Google Earth. Scale bar = 10 km

Fig. 6.45 Dayas, Hamada du Guir, Algeria. 31°20′4.47″N, 3°26′6.71″W. ©Google Earth. Scale bar = 700 m

Fig. 6.46 Kharga, Egypt, wind-fluted limestone. 25°47′33.59″N, 30°20′7.56″E. ©Google Earth. Scale bar = 3 km

this hyper-arid area receives less than 1 mm of annual precipitation. Smith et al. (2004a, 2004b) believed that the spring tufas that occur along the edge of the plateau record spring activity at relatively high elevations during times of increased rainfall. The tufas formed either where surface streams draining the plateau cascaded over the escarpment, or where spring waters issued from beneath the plateau's limestone capping. Direct dates on spring carbonates and other pluvial phase sediments suggest that spring activity occurred intermittently between 100 and 350 ka, with some evidence of a minor rainfall event at ~45 ka.

Elsewhere in the Libyan Desert, at Bahariya (Fig. 6.48) and Farafra (Fig. 6.49), there are areas with residual hills, caves and poljes that may be fossil karst related to even earlier former pluvial conditions (Embabi 2018; El-Aref et al. 1987) in post-Miocene times.

6.10 Calcrete

In dry areas, limestone deposits called calcrete occur. Calcrete is a term a term that was originally proposed by Lamplugh (1902), to describe a terrestrial near-surface accumulation of predominantly calcium carbonate ($CaCO_3$) which occurs in a variety of forms ranging from powdery to nodular to highly indurated. It results from low temperature physico-chemical processes operating within the zone of weathering which lead to the displacive and/or replacive introduction of $CaCO_3$ into a soil profile, sediment, rock or weathered material. They develop as a result of carbonates in solution moving laterally and vertically through vadose and shallow phreatic groundwater systems until they become, over time, saturated with respect to $CaCO_3$ and precipitate as calcite crystals. In the USA, the term is synonymous with *caliche*. Calcretes are estimated to underlie 13% of the Earth's land surface and are most widespread in semi-arid regions. Important areas of occurrence include the High Plains of the USA, Africa north of the Sahara, the Kalahari of southern Africa, central and western Australia, and parts of southern Europe. Their origin and characteristics have been described by Goudie (1973), Reeves (1976) and Wright (2009). One of their roles is to add resistance to sediment accumulations (e.g. river terrace gravels) and to provide a protective cap to many landforms.

Superb examples of calcrete caprocks are present above the Kuiseb Valley in Central Namibia (Goudie and Viles 2015a, b) (Fig. 6.50). Here the Great Escarpment, into which the Kuiseb is incised, is capped by the Karpenkliff Conglomerate (Ritter et al. 2023). This consists of gravel deposits, which probably had an alluvial fan origin, which

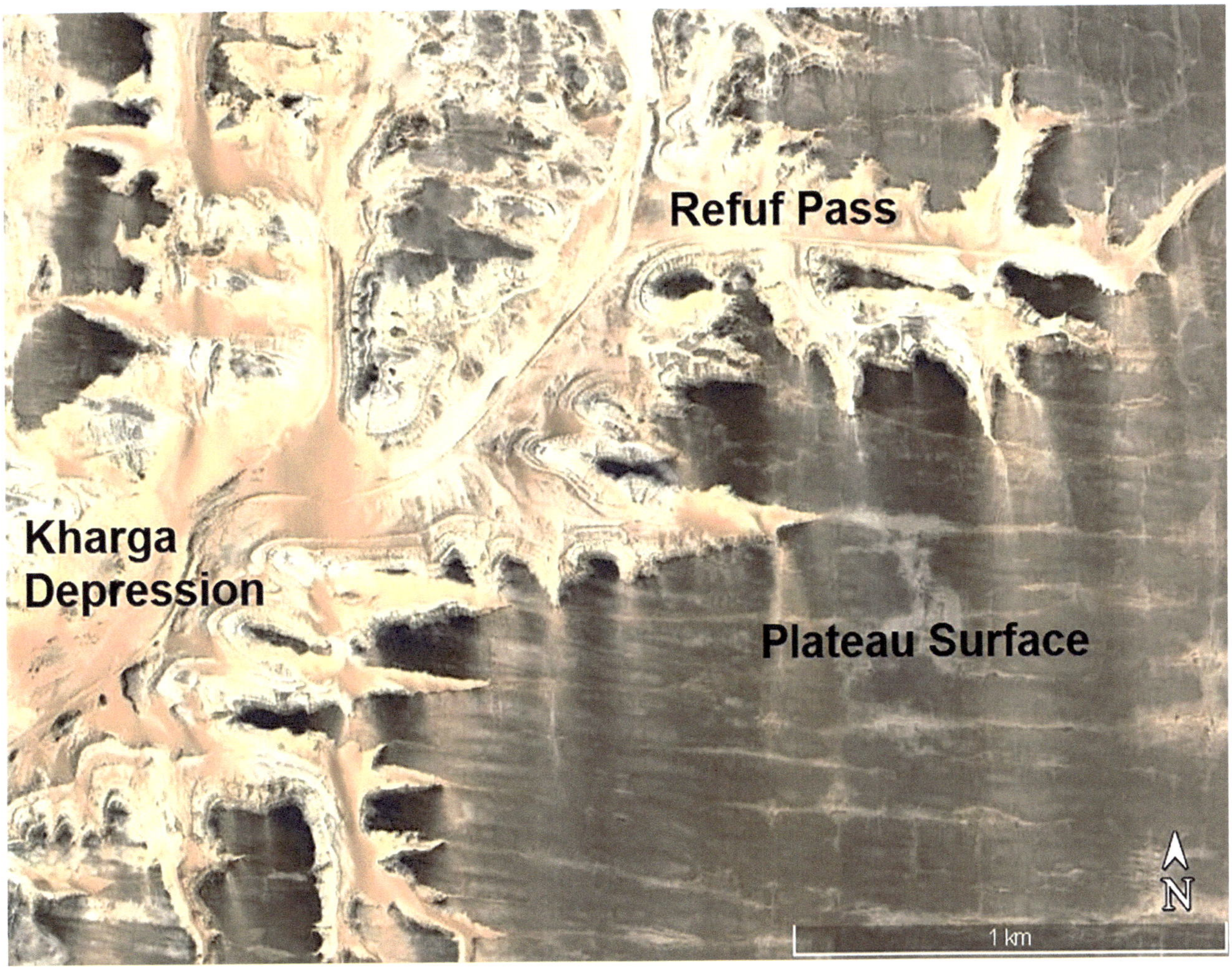

Fig. 6.47 Refuf Pass, Egypt, groundwater sapping of limestones. 25°41′13.77″N, 30°50′22.58″E. ©Google Earth. Scale bar = 1 km

have become cemented by calcrete The calcrete, which is up to 5 m thick, is believed to pre-date the incision of the Kuisbeb Canyon and may date back towards the end of the Miocene. There is a marked contrast between the calcrete capped surface, with its limited drainage development and with the occurrence of small solutional depressions (dayas), and the gullied schist bedrock below. The continued existence of the calcrete reflects increasing Namibian aridity since the Miocene, with its excellent preservation implying that arid conditions may have been maintained to the present. The incision and the consequent preservation of the capped surfaces have been attributed to late Tertiary/Quaternary epeirogeny.

Other calcrete-capped surfaces occur in southern Africa, as along the Molopo River on the borders of Botswana and South Africa (Fig. 6.51). The terrace remnants of the Ugab River in central Namibia also show a beautiful calcrete capping (Fig. 6.52) (https://www.mme.gov.na/files/publications/15c_A4_Vingerklip_en.pdf) (Accessed 30th August 2024). Here there is a main terrace which occurs at c 160 m above the current level of the Ugab River and which consists of a fluvial conglomerate cemented with carbonate (Schneider 2008, Sect. 8.31) some of which is translocated from nearby dolomite outcrops.

In the USA, the High Plains (Llano Estacado) are frequently underlain by calcretes (caliches) of the Ogallala Formation, and the surface of the plains contrasts strongly with the serrated and gullied escarpment face. It is also notable for the formation of small closed depressions—*playas* or pans—that may in part have a solutional origin (Reeves 1970, 1971) (Fig. 6.53). Some of the High Plains calcrete may be associated with inverted relief, with the calcrete originally having formed along drainage lines and now being etched out by subsequent erosion of the non-cemented beds within the Ogallala (Reeves1983). The whole question of the relationships between calcrete development, pan distribution, pseudo-karst and patterned ground in the USA and southern Africa is given in Goudie (1973, pp. 52–65).

(a)

(b)

Fig. 6.48 **a** Bahariya cones, Egypt. 28°22′5.27″N, 29°11′5.34″E. ©Google Earth. Scale bar = 300 m. **b** Bahariya yardangs, Egypt, with cones in background and yardangs ('mud lions') in the foreground

(a)

(b)

Fig. 6.49 **a** Farafra cones. 27°20′3.98″N, 28°3′20.96″E. ©Google Earth. Scale bar = 1 km. **b** Farafra cones (Photo A.S. Goudie)

(a)

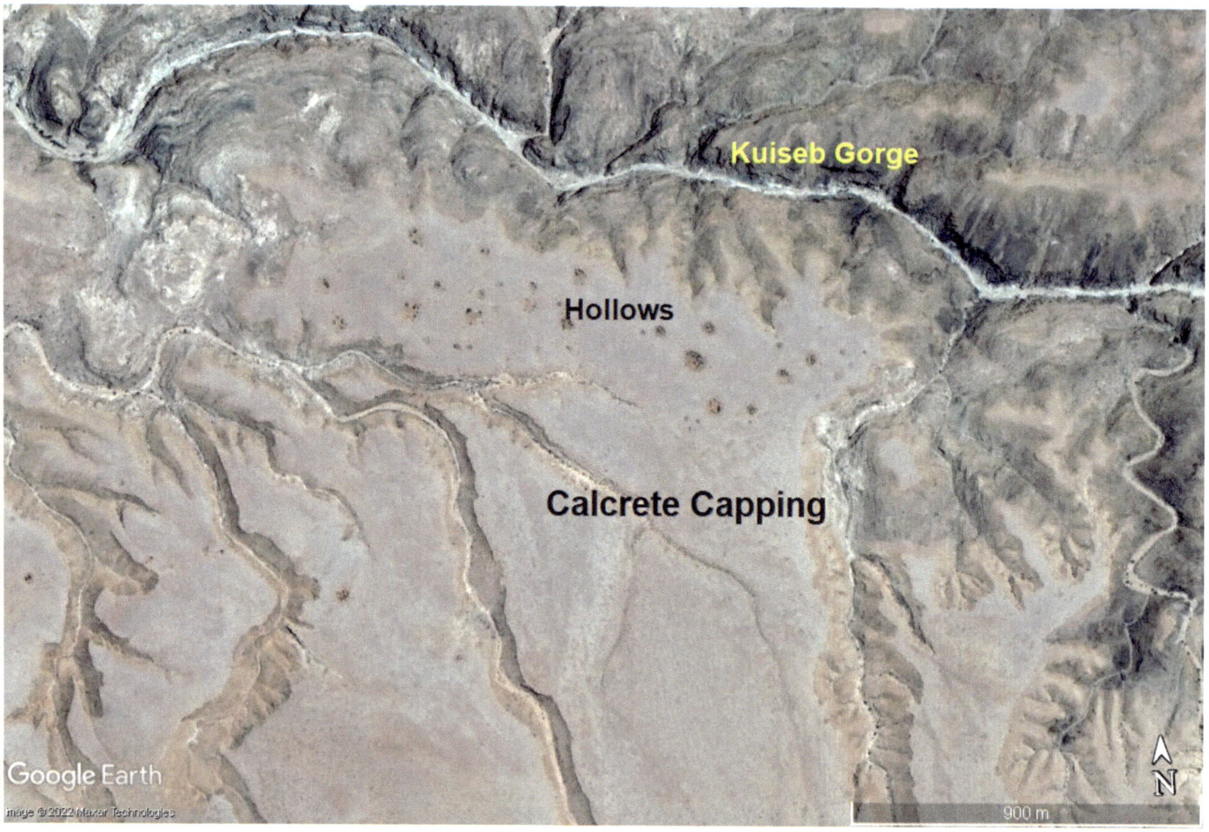

(b)

Fig. 6.50 **a** Kuiseb valley, Namibia, showing calcrete capping and small solution hollows. 23°29'S, 15°48'E. ©Google Earth. Scale bar = 900 m. **b** Karpenkliff, Namibia, with calcreted conglomerate caprock (Photo A.S. Goudie)

(a)

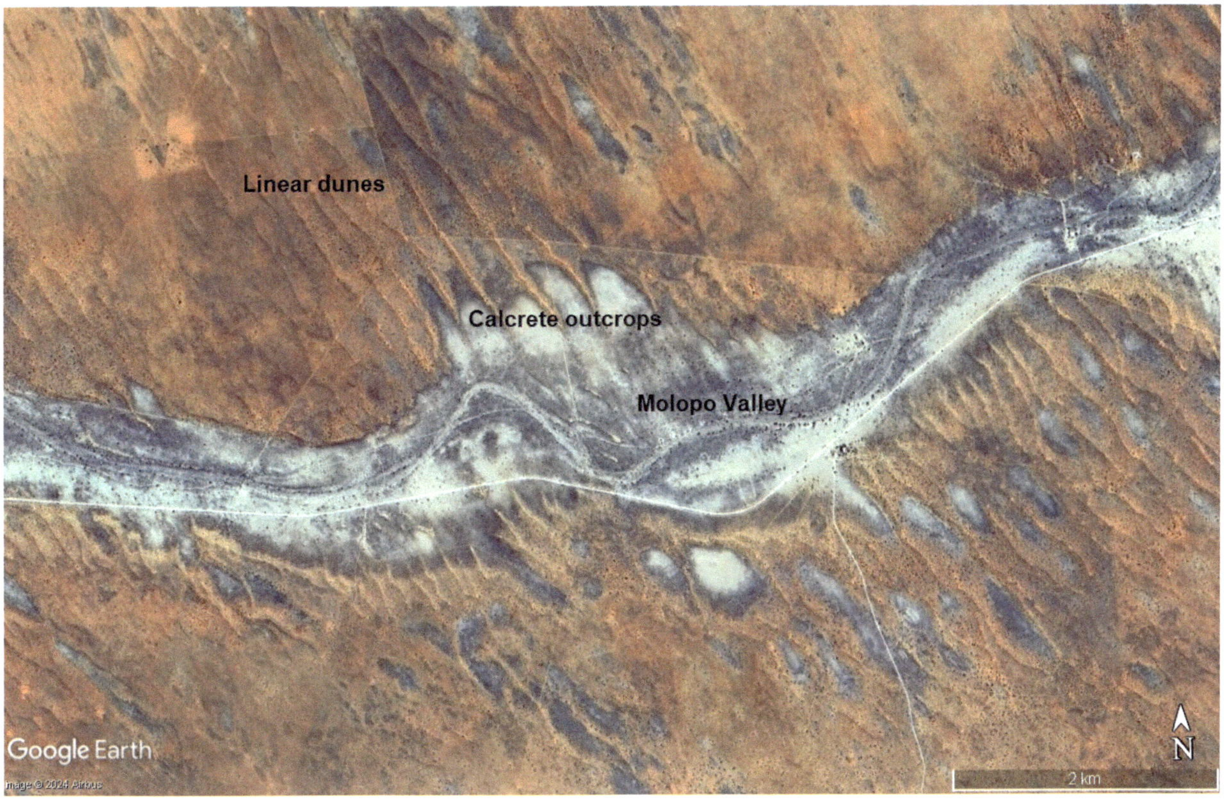

(b)

Fig. 6.51 **a** Molopo River Calcrete, Botswana/South Africa. 27°4′3.41″S, 21°32′45.72″E. ©Google Earth. Scale bar = 2 km. **b** Molopo River, South Africa and Botswana, with calcrete capping (Photo A.S. Goudie). **c** Calcrete caprock lining sides of Molopo Valley, South Africa and Botswana (Photo A.S. Goudie)

(c)

Fig. 6.51 (continued)

Fig. 6.52 Calcrete-capped plateau, Ugab Valley, Namibia. 20° 24'S, 15°25'E. ©Google Earth. Scale bar = 200 m

(a)

(b)

Fig. 6.53 **a** Calcrete (caliche) capped Llano Estacado in Texas, USA. 33°33′54.52″N, 101°12′17.12″W. ©Google Earth. Scale bar = 6 km. **b** Calcrete (caliche) capping, Llano Estacado, Texas, (Photo A.S. Goudie). C.C. 'Tex' Reeves for scale

Fig. 6.54 Inverted channels in Oman. 22°17′32.48″N, 57°49′18.18″E. ©Google Earth. Scale bar = 2 km

The cementation of river channel deposits by calcrete has also contributed to relief inversion in Oman to the west of the Wahiba Sands. Here sinuous gravel trains, laid down by streams issuing from the Eastern Oman Mountains, have been left upstanding by 10–20 m above the adjacent plains (Maizels 1987) (Fig. 6.54). Similar inverted refief, sometimes called suspendritic drainage or drainage in bas relief, is also found in other parts of Arabia (Miller 1937).

6.11 Spring Mounds

In some drylands, carbonate deposits, often described as 'pinnacles' were formed under pluvial conditions when the material was deposited by springs that occurred on the floors of lakes. The lake basins of North America are particularly notable for their tufa pinnacles, as in Searles Lake, Lake Tecopa (Nelson et al. 2001), Mono Lake (Dunn 1953), Owens Lake (Bradbury 1997), and Lake Lahontan (Demott et al. 2019). Tufa pinnacles on the floors of lakes also occur in the Badain Jaran desert of China (Arp et al. 1998).

Not all tufa mounds have developed under lakes—they can also occur as a result of springs discharging at the ground surface in locations like the Great Artesian Basin of Australia (Inverarity et al. 2016), Tunisia (Roberts and Mitchell 1987) or the depressions of the Libyan Desert, such as Dakhla (Adelsberger and Smith 2010).

Searles Lake is a dry lake within the Searles Basin, a north-trending graben of the Basin-and-Range Province in southeastern California (Rieger 1992). Over 500 tufa mounds or pinnacles occur in its dry bed. Most are 5–12 m high but some reach heights of as much as 45 m (Fig. 6.55). They are composed of calcite and aragonite, and formed in association with spring vents in the Pleistocene lake bottom. The maximum lake depth was up to 200 m. Ground water flowed into the lake through fractures associated with the Garlock Fault (Guo and Chafetz 2012) and its was from spring orifices that this groundwater deposited the tufa. There are no active springs in the area now. Scholl (1960, p. 414) described the pinnacles as 'tower-, tombstone-, and cone-shaped masses and large limestone ridges.' He believed that deposition of the pinnacles at the south west end took place in a lake which filled Searles basin more than 32,000 years ago. Pinnacles midway along and at the northern end formed in a late Pleistocene lake which flooded the basin between 23,000 and 10,000 years ago.

Spring mounds are widespread in North Africa. Those in southern Tunisia are 25–30 m high and occur in great profusion on the south side of the Chott el Djerid (Roberts and Mitchell 1987). There are also many tufas and spring

Fig. 6.55 **a** Trona Pinnacles, Searles Lake, California. 35°36′56.02″N, 117°22′20.59″W ©Google Earth. Scale bar = 90 m. **b** Trona Pinnacles, Searles Lake, California (Photo A.S. Goudie)

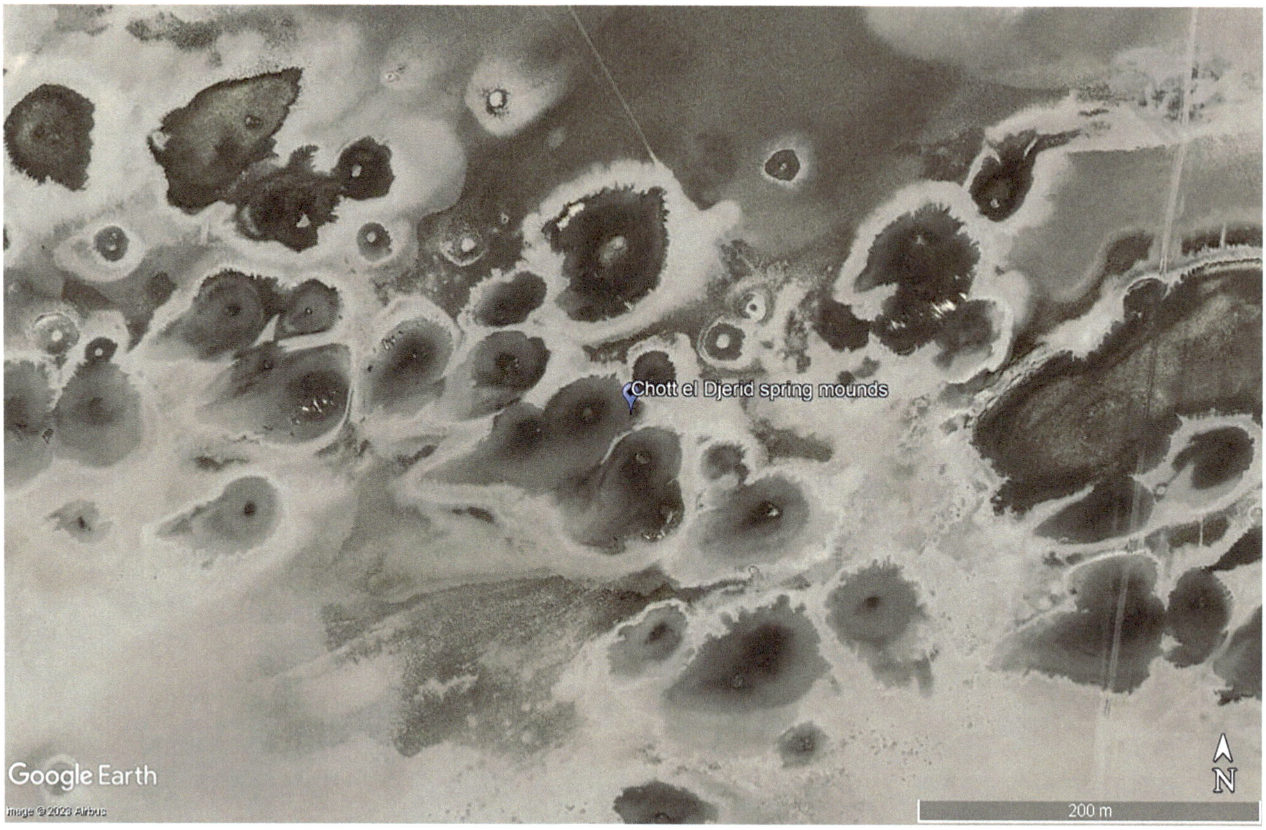

Fig. 6.56 Chott el Djerid spring mounds, Tunisia. 33°38′26.85″N, 8°39′7.49″E. ©Google Earth. Scale bar = 200 m

Fig. 6.57 Blasskrantz Tufa, Namibia. The tufa mound in the centre of the image is c 80 m high and 400 m across (Photo A.S. Goudie)

Fig. 6.58 Tufa mound at foot of Devonian reef cliffs, Napier Range, Australia (Photo A.S. Goudie)

Fig. 6.59 Pamukkale travertine terraces, Turkey. 37°55′31.84″N, 29°7′18.36″E. ©Google Earth. Scale bar = 50 m

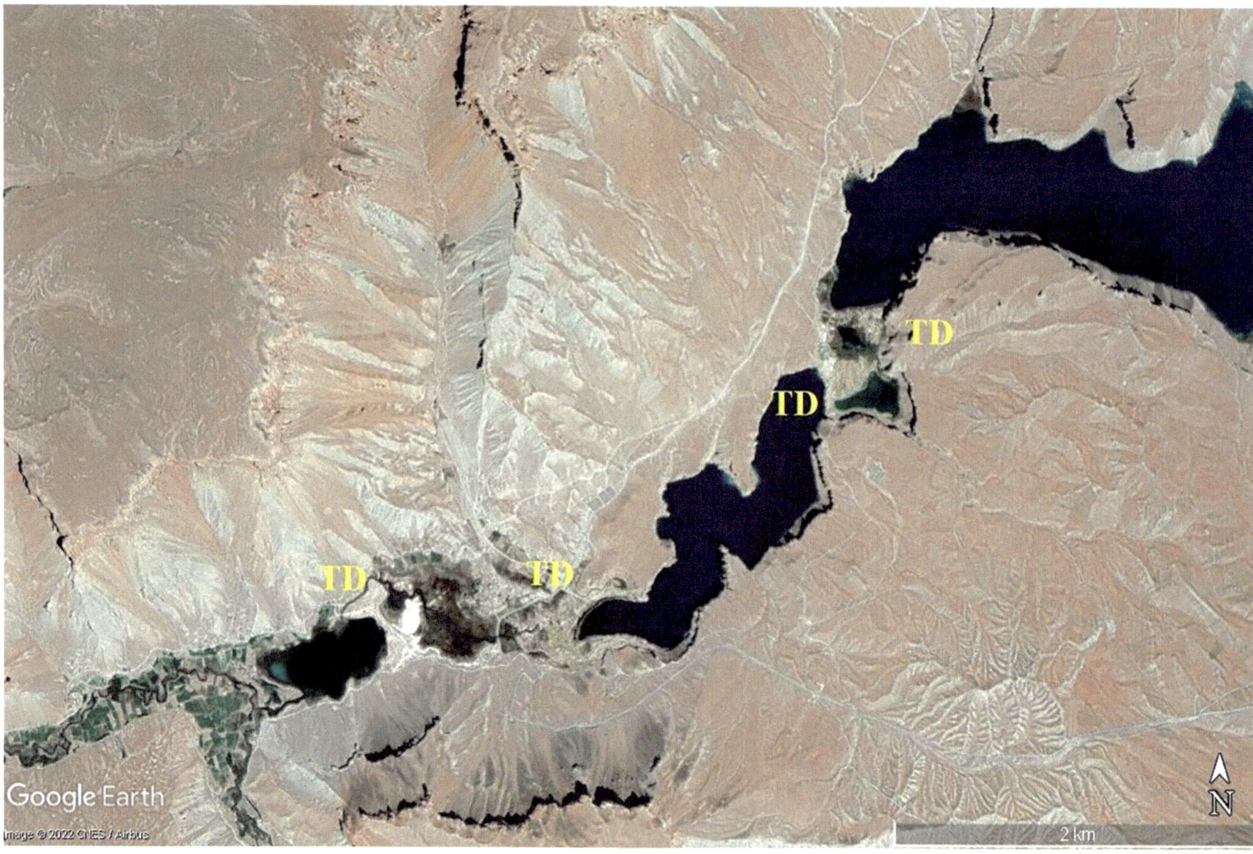

Fig. 6.60 Band-e-Amir lakes and tufa dams (TD), Afghanistan. 34°50'N, 67°11'E. ©Google Earth. Scale bar = 2 km

mounds in the Libyan Desert of Egypt, and these may date back to Pleistocene pluvial times (Crombie et al. 1997) (Fig. 6.56).

6.12 Tufas and Travertines

Tufas are freshwater carbonate deposits which are widespread in desert areas with carbonate rocks, but also in many other climatic zones (Viles and Pentecost 2007). These are formed from ambient, non-thermal waters. Some of these are deposited in fluvial channels and form a series of dams or barrages (e.g. Band-e-Amir in Afghanistan), and some are deposited as cascades as streams debouch over mountain fronts (e.g. at Blasskrantz in the Naukluft Mountains of Namibia) (Viles et al. 2007) (Fig. 6.57).

Others are deposited by springs issuing on slopes, others are formed as algal bioherms on the margins of lakes, and others are laid down in swamps and other poorly drained areas (paludal tufas). Massive quantities of tufa, with a great range of morphologies occur in the semi-arid and subhumid Oscar and Napier Ranges of the Kimberley District of north west Australia (Viles and Goudie 1990; Wright 2000): drapes formed on cliff faces, cones where water issues from caves, barriers in channels, and rimstone pools on gentle pediments (Fig. 6.58). Some of the older tufas have been karstified (to produce caves and karren) or have been altered into calcrete crusts. The intense seasonality of the rainfall possibly facilitates carbonate solution and deposition.

Some tufas (travertines) are associated with deposition from thermal waters, famous examples of which are found at the World Heritage Site of Pammukale in south western Turkey (Altunel and Hancock 1993a, b). This may be the largest travertine deposit in the world (https://earthobservatory.nasa.gov/images/149127/turkeys-cotton-castle) (accessed 25th June 2024). Here travertine deposition has been in progress for at least the last 400,000 years, and has partially overwhelmed the Roman city and necropolis of Hierapolis. It forms white terraces and perched channels of wall-like form that are self-built as a result of man channelling waterflow The travertine originates from hot waters that emerge from underlying limestone bedrock at 35–56°C via open fissures and at least one fault zone (Fig. 6.59). Pentecost (2005) provides a fine review of global travertines.

Band-e-Amir is a group of six beautiful lakes that make up Afghanistan's first national park. They lie in the Hindu

Fig. 6.61 Tufa dams and blue lakes, Wadi Darbat, Oman. 17°4′40.22″N, 54°26′12.77″E. ©Google Earth. Scale bar = 400 m

Kush Mountains at an elevation of about 3000 m, some 200 km west of Kabul. Their waters are noted for their brilliant emerald colour (Fig. 6.60). They have formed in a broad valley in a semi-arid region that is fed with water from melting snow which percolates through the limestone bedrock of the area, producing perennial springs (Bourrouilh-Le Jan et al. 2007; Lang and Lucas 1970). The lakes are impounded by the growth of a series of tufa dams that are believed to have formed in the Holocene. These form barrages that extend for some 12 km along the valley and have nearly vertical walls some 10 m high.

In southern Oman, near Salalah, the Jebel Qara massif is a limestone plateau deeply cut by an intricate system of dry valleys (wadis) mostly draining into the Arabian Sea. It is formed by Late Cretaceous to Neogene limestones and comprises a steep escarpment abutting a coastal plain. One of these wadis is the Wadi Darbat. This flows south from the Dhofar plateau out into Khor Rori. Inland from Khor Rori, the wadi shows an impressive travertine curtain, a petrified waterfall over 150 m high. According to Searle (2019a, b, Chap. 15), 'during the *khareef* floods Wadi Darbat in full spate is one of the most incredible sights in Arabia, Oman's answer to the Victoria Falls'. The deposition of tufa appears to have occurred throughout the Quaternary. Tufa dams act as barrages along wadis, forming tufa-dammed lakes. Some reach remarkable dimensions: one of the largest, closing the Wadi Darbat, forms a c 130 m high and c 700 m long cliff (Zerboni et al. 2020) (Fig. 6.61).

The Plitvice lakes in Croatia are some of the finest examples of lake impoundment by tufa dams. Sixteen lakes have formed along a 6.5 km stretch of the upper Korana valley. The tufa barriers reach 30 m high, and compound dams hold back lakes as deep as 46 m (Fig. 6.62). Full details of their geomorphological and hydrological characteristics are given by Radišić et al. (2021).

High altitude tufa dams occur in China at Jiuzhaigou (Fig. 6.63), which is situated in the Min Shan Mountains at the eastern margin of the Tibetan Plateau, north of the Sichuan basin. The bedrock in the area consists of carbonate rocks of Devonian through Triassic age (Florsheim et al. 2013; Lv et al. 2023) and the area's rainfall is highly seasonal. The dams, which occur in the Shuzheng valley, date back to the early Holocene, and were badly affected by a major earthquake in 2017. Other beautiful tufa-dammed lakes occur in the Huanglong valley, which is situated further south in the north-west of Sichuan Province (32°44′27.31″N, 103°49′51.32″E) (Lu et al. 2000).

Fig. 6.62 Plitvice lakes, Croatia. 44°52′26.98″N, 15°36′20.22″E. ©Google Earth. Scale bar = 500 m

Fig. 6.63 A string of over 20 tufa dams in the Shuzheng valley, China. 33°12′13.48″N, 103°54′5.92″E. ©Google Earth. Scale bar = 400 m

(a)

(b)

Fig. 6.64 **a** Aeolianite ridges (A) in the Wahiba Sands, Oman. 21°9′34.37″N, 58°55′35.77″E. ©Google Earth. Scale bar = 500 m. **b** Aeolianite ridges between linear dunes in the Wahiba Sands. Note the cross-bedding structures (Photo A.S. Goudie)

Fig. 6.65 Shoal Cape, Western Australia. 33°52′7.00″S, 121°10′15.27″E. ©Google Earth. Scale bar = 200 m. Here there are three beachrock ridges showing the progress of beach retreat. The beachrock is more resistant than unlithified beach materials. It also provides some protection against wave erosion

6.13 Coastal Limestones—Aeolianite, Beachrock and Raised Reefs

As we saw in the case of the Great Barrier Reef (Sect. 6.7), some limestone rocks are forming today in coastal environments and are creating characteristic landscapes. Brief mention also needs to be made of three other rock types from coastal environments—aeolianites, beachrock, and raised reefs.

Aeolianites are carbonate-cemented dune sands (Brooke 2001). They are found in a wide variety of climatic zones. A number of tropical islands are made up largely of aeolianite, as for example in Bermuda and the Bahamas, and very extensive deposits occur in South Africa and Australia (McLaren 2007). In India they are termed *miliolites* (Biswas 1971) and are rich in the tests of Foraminifera (Sperling and Goudie 1975). Particularly fine examples of aeolianite occur in the Wahiba Sands of Oman (Goudie et al. 1987) (Fig. 6.64).

In inter-tidal areas, including in the Mediterranean basin (Goudie 1969), beach sediments may be lithified to form beachrock (Stoddart and Cann 1965) (Fig. 6.65).

In some tropical regions, there may be large spreads of supra-tidal limestones composed of elevated reefs, some of which date back to the high sea levels of the last interglacial. This is the case, for instance, in Barbados (Schellmann and Radke 2004). There are also various other carbonate constructions created by organisms such as vermetids, but these are generaly too small to show up on Google Earth. Upraised reefs, such as at Christmas Island (Grimes 2001) and Aldabra Atoll (Trudgill 1979) in the Indian Ocean, display a wide range of karstic forms.

Rather older raised reefs accur in the Cook Islands in the Pacific. A notable example is the island of Mangaia. This originated over a hot spot at c 17—18 mya. The centre of the island is occupied by a basaltic volcano, rising to c 167 m above sea level. The next zone is a low

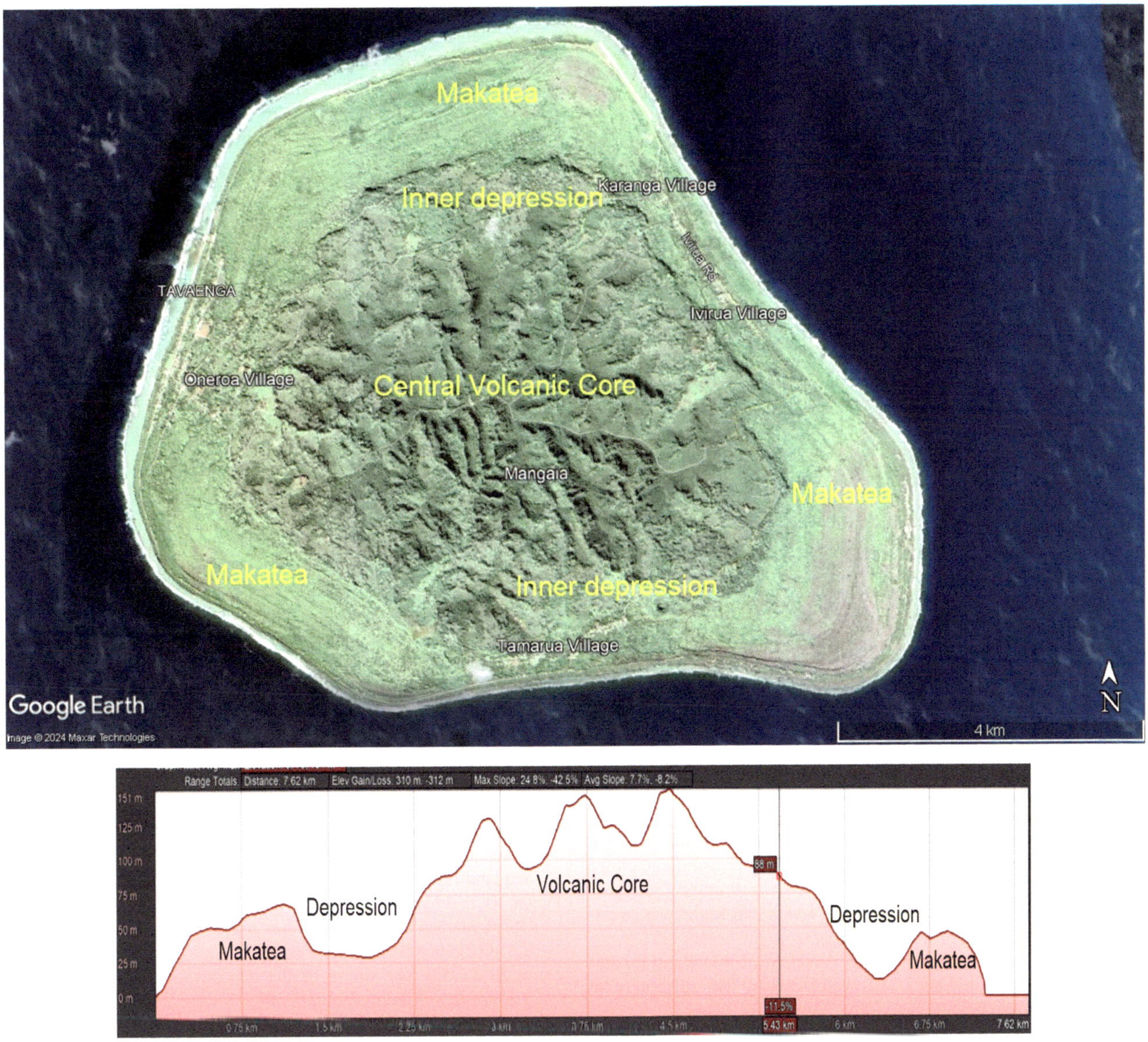

Fig. 6.66 **a** Mangaia in the Cook Islands. 21°55′31.76″S, 157°54′44.24″W. ©Google Earth. Scale bar = 4 km. Lower diagram is approximate north–south elevation cross-profile

lying, swampy area or depression, rising only a few metres or tens of metres above sea level. Further seaward is a steep fringing wall, upto 55–70 m high and 500–1000 m wide, of ancient coral called *makatea*. The *makatea* is an emerged fringing reef, which has been steepened on its inland side by solutional attack caused by undersaturated water running off the volcanics. The depression is not an old lagoon that formed inland from a barrier reef, but is an essentially karstic feature (Stoddart et al. 1985). There are limestone outliers found on the volcanics, which imply that the *makatea* was indeed once continuous across the depression (Fig. 6.66). The *makatea* is riddled with caves (Ellison 1994) and pocked with karren.

6.14 Conclusions

Karstic landforms associated with soluble rocks (e.g. limestone, dolomite, gypsum, anhydrite, and halite) are among the most striking in the world. They include a range of features including gorges, underground drainage, closed depressions, flutes, cones and towers. The nature of these rocks (solubility, porosity, permeability, jointing, bedding, etc.) contributes to the diversity of forms that is observed, but also significant is climate, past and present. Although solutional features predominate, in some environments one also has depositional forms, such as tufas, travertines, and spring mounds. In coastal environments one may find aeolianites, beachrock and raised reefs.

References

Abotalib AZ, Sultan M, Elkadiri R (2016) Groundwater processes in Saharan Africa: implications for landscape evolution in arid environments. Earth Sci Rev 156:108–136

Adelsberger KA, Smith JR (2010) Paleolandscape and paleoenvironmental interpretation of spring-deposited sediments in Dakhleh Oasis, Western desert of Egypt. CATENA 83(1):7–22

Allison RJ, Goudie AS (1990) The form of rock slopes in tropical limestone and their associations with rock mass strength. Z Geomorph 34(2):129–148

Allison RJ, Goudie AS, Cox NJ (1993) Geotechnical properties of rock masses: their control on slope form and mechanisms of change along the Napier Range, Western Australia. Geomorphology 8(1):65–80

Altunel ER, Hancock PL (1993a) Morphology and structural setting of quaternary travertines at pamukkale, Turkey. Geol J 28(3–4):335–346

Altunel ER, Hancock PL (1993b) Active fissuring and faulting in quaternary travertines at pamukkale, western Turkey. Z Geomorph Suppl 94:285–302

Andreychouk V, Eftimi R, Nita J, Klimchouk A (2021) Geomorphology and hydrogeology of an exposed evaporite dome: the Dumre karst area, Central Albania. Geol Quart 65(4):55–65

Arp G, Hofmann J, Reitner J (1998) Microbial fabric formation in spring mounds ("microbialites") of alkaline salt lakes in the Badain Jaran Sand Sea, PR China. Palaios 13:581–592

Baadi K, Lebreton B, Pereira P (2023) Hydric geosites of middle atlas: a natural heritage under threat. In: Baadi K (ed) Geoheritage of the middle atlas (Morocco). Geoheritage, geoparks and geotourism. Springer, Cham. https://doi.org/10.1007/978-3-031-27073-4_8

Baechler F, Boehner R (2014) Karst geology and hydrogeology of Cape Breton Island, Nova Scotia: an overview. Can J Earth Sci 51(7):701–714

Beck BF, Sinclair WC (1986) Sinkholes in Florida: an introduction. Univ Central Florida, Florida Sinkhole Research Institute, College of Engineering, p 16

Biswas SK (1971) The miliolite rocks of Kutch and Kathiawar (western India). Sed Geol 5(2):147–164

Bourrouilh-Le Jan FG, Akram B, Schvoerer M (2007) Band-e-Amir Lakes and Dragon Valley (Bamiyan): myths and seismicity in Afghanistan. Geol Soc London Spec Pub 273(1):121–132

Bradbury JP (1997) A diatom record of climate and hydrology for the past 200 ka from Owens Lake, California with comparison to other Great Basin records. Quat Sci Rev 16:203–219

Brook GA, Ford DC (1978) The origin of labyrinth and tower karst and the climatic conditions necessary for their development. Nature 275:493–496

Brooke B (2001) The distribution of carbonate eolianite. Earth Sci Rev 55(1–2):135–164

Calaforra JM, Pulido-Bosch A (1999) Gypsum karst features as evidence of diapiric processes in the Betic Cordillera, Southern Spain. Geomorphology 29(3–4):251–264

Ćalić J (2011) Karstic uvala revisited: toward a redefinition of the term. Geomorphology 134(1–2):32–42

Calleja I, Tonelli C (2019) Dwejra and Maqluba: emblematic sinkholes in the Maltese Islands. In: Gauch R, Schembri JR (eds) Landscapes and landforms of the Maltese Islands. Springer, Cham, pp 129–139

Cardinal T et al (2024) The interplay of geology, climate and tectonics on river incision: the example of the High Verdon Gorges, southwestern French Alps. CR Géosci 356(S2):1–23

Castellani V, Dragoni W (1986) Evidence for karstic mechanisms involved in the evolution of Moroccan hamadas. Int J Speleol 15(1):57–71

Cerrano C et al (2006) Marine lakes of karst islands in Ha Long Bay (Vietnam). Chem Ecol 22(6):489–500

Chardonnet J (1948) Les Calanques Provençales. Origine Et Divers Type. Ann Géogr 57:289–297

Corbel J (1956) Un karst Méditerranéen de basse altitude: le massif des Calanques et la formation de son relief. Géocarrefour 31(2):129–136

Crombie MK, Arvidson RE, Sturchio NC, El Alfy Z, Zeid KA (1997) Age and isotopic constraints on Pleistocene pluvial episodes in the Western Desert, Egypt. Palaeogeogr Palaeoclimatol Palaeoecol 130(1–4):337–355

Cvijić J (1901) Morphologische und glaciale Studien aus Bosien, Herzegovina und Montenegro. Abhand Geogr Gesell 3:1–85

Daoxian Y (1987) New observations on tower karst. In: Gardiner V (ed) International geomorphology 1986, part 2. Wiley, Chichester, pp 1109–1123

Day MJ (1978) Morphology and distribution of residual limestone hills (mogotes) in the karst of northern Puerto Rico. Geol Soc Am Bull 89(3):426–432

Day MJ (2004) Cockpit Country cone karst, Jamaica. In: Gunn J (ed) Encyclopedia of caves and karst science. Fitzroy Dearborn, New York, pp 233–235

de la Rosa JPM (2016) The Burren: a glacial, karstic and biokarstic expression of a limestone plateau in western Ireland. Earth Surf Proc Landf 41(11):1614–1628

De Waele J, Melis MT (2009) Geomorphology and geomorphological heritage of the Ifrane-Azrou region (Middle Atlas, Morocco). Environ Geol 58:587–599

De Waele J, Picotti V, Zini L, Cucchi F, Forti PA (2009) Karst phenomena in the Cordillera De La Sal (Atacama, Chile). GeoActa Special Pub 2:113–127

Demott LM, Scholz CA, Junium CK (2019) 8200-year growth history of a Lahontan-age lacustrine tufa deposit. Sedimentology 66(6):2169–2190

Doğan U, Yeşilyurt S (2019) Gypsum karst landscape in the Sivas Basin. In: Kuzucuoğlu C, Ciner A, Kazanci N (eds) Landscapes and landforms of Turkey. Springer, Cham, pp 197–218

Doğan U, Yılmaz M (2011) Natural and induced sinkholes of the Obruk plateau and Karapınar-hotamış plain, Turkey. J Asian Earth Sci 40(2):496–508

Dunn JR (1953) The origin of the deposits of tufa in Mono Lake (California). J Sedim Res 23(1):18–23

Eberhard RS, Sharples C (2013) Appropriate terminology for karst-like phenomena: the problem with 'pseudokarst.' Int J Speleol 42(2):109–113

El-Aref MM, Abou Khadrah AM, Lotfy ZH (1987) Karst topography and karstification processes in the Eocene limestone plateau of El Bahariya Oasis, Western Desert, Egypt. Z Geomorph NF31:45–64

Ellison JC (1994) Caves and speleogenesis of Mangaia, Cook Islands. Atoll Res Bull 417:1–25

Elshaafi A, Bilal A, Awami Y, Salloum F (2021) Development of collapse dolines in Al Jabal Al Akhdar, NE Libya: geomorphic analysis and numerical modeling-based approach. *Carbonates and Evaporites*, 36(3):54

Embabi NS (ed) (2018) Landscapes and landforms of Egypt. Springer, Cham, p 336

Florsheim JL et al (2013) Basin-scale and travertine dam-scale controls on fluvial travertine, Jiuzhaigou, southwestern China. Geomorphology 180:267–280

Ford TD (ed) (1977) Limestones and caves of the Peak District. Geoabstracts, Norwich, p 469

Ford TD (ed) (1989) Limestones and caves of Wales. Cambridge Univ Press, Cambridge, p 259

Ford DC (2006) Jovan Cvijić and the founding of karst geomorphology. Environ Geol 51:675–684

Ford DC (2010) The South Nahanni: high-latitude limestone landscapes. In: Migon P (ed) Geomorphological landscapes of the World. Springer, Dordrecht, pp 13–20

Ford DC, Williams PW (2007) Karst hydrology and geomorphology (2nd edn). Wiley, Chichester, p 526

Ford DC (2016) The many landscapes of South Nahanni National Park. In: Slaymaker O (ed) Landscapes and landforms of Western Canada. Springer, Cham p 435

Gabrovšek F, Stepišnik U (2011) On the formation of collapse dolines: a modelling perspective. *Geomorphology*, 134(1–2):23–31

Garasic M (2012) Crveno Jezero—the biggest sinkhole in Dinaric Karst (Croatia). In Abbasi A, Giesen N (eds), EGU General Assembly Conference Abstracts 14:7132

Gatt P (2019) Evaporite dissolution sinkholes in the Dwejra Depression, Malta. Eur Geol 48:53–57

Gillieson DS (2021) Caves: processes, development, and management (2nd edn). Wiley-Blackwell, Chichester, p 528

Ginés A, Knez M, SlabeT, Dreybrodt W (eds) (2009) Karst rock features. Karren Sculpturing. Založba ZRC, p 561

Gischler E et al (2017) A giant underwater, encrusted stalactite from the Blue Hole, Lighthouse Reef, Belize, revisited: a complex history of biologically induced carbonate accretion under changing meteoric and marine conditions. J Sed Res 87(12):1260–1284

Gökkaya E, Gutiérrez F (2022) Poljes in the Sivas gypsum karst, Turk ey. Geomorphology 417:108451

Goldscheider N et al (2020) Global distribution of carbonate rocks and karst water resources. Hydrogeol J 28:1661–1677

Gómez-Pujol L et al (2013) Structural control on the presence and character of calas: observations from Balearic Islands limestone rock coast macroforms. Geomorphology 194:1–5

Goudie AS (1969) A note on Mediterranean beachrock: its history. Atoll Res Bull 126:6–14

Goudie AS (1973) Duricrusts of tropical and sub-tropical landscapes. Clarendon Press, Oxford, p 192

Goudie AS (1999) A comparison of the relative resistance of limestones to salt and frost weathering. Permafr Periglac Proc 10:309–316

Goudie AS (2010) Dayas: distribution and morphology of dryland solutional depressions developed in limestones. Z Geomorph 54:145–159

Goudie AS, Gardner RAM (1985) Discovering landscape in England and Wales. Allen and Unwin, London, p 177

Goudie AS, Viles H (2015a) Calcretes: the Kamberg Calcrete Formation and the Karpenkliff Conglomerate. In: Goudie AS, Viles H (eds) Landscapes and landforms of Namibia. Springer, Dordrecht, pp 111–114

Goudie AS, Viles H (2015b) The Weissrand and the aligned drainage of the western Kalahari. In: Goudie AS, Viles H (eds) Landscapes and landforms of Namibia. Springer, Dordrecht, pp 143–146

Goudie AS, Warren A, Jones DKC, Cooke RU (1987) The character and possible origins of the aeolian sediments of the Wahiba Sand Sea, Oman. Geogr J 153:231–256

Goudie A, Viles H, Allison R, Day M, Livingstone I, Bull P (1990) The geomorphology of the Napier Range, Western Australia. Trans Inst Brit Geogr 15:308–322

Grimes KG (2001) Karst features of Christmas Island (Indian Ocean). Helictite 37(2):41–58

Grimes KG (2012) Karst and paleokarst features involving sandstones of the Judbarra/Gregory National Park, Northern Territory, Australia. Helictite 41:67–73

Grimes KG, Spate A (2008) Laterite Karst. Ackma J 73:49–52

Guo X, Chafetz HS (2012) Large tufa mounds, Searles Lake, California. Sedimentology 59(5):1509–1535

Gutiérrez F, Guerrero J, Lucha P (2008) A genetic classification of sinkholes illustrated from evaporite paleokarst exposures in Spain. Environ Geol 53:993–1006

Haryono E, Day M (2004) Landform differentiation within the Gunung Kidul Kegelkarst, Java, Indonesia. J Cave Karst Studies 66(2):62–69

He C (1987) The characteristics of karst geomorphology in Guizhou. In: Gardiner V (ed) International geomorphology 1986, part 2. Wiley, Chichester, pp 1095–1108

Hennessy R, McNamara M, Hoctor Z (2017) Stone, water and ice (3rd edn). Clare County Council, Ennistymon, p 132

Hopley D, Smithers SG, Parnell K (2007) The geomorphology of the Great Barrier Reef: development, diversity and change. Cambridge Univ Press, Cambridge, p 470

Inverarity K, Heinson G, Hatch M (2016) Groundwater flow underneath mound spring tufas from geophysical surveys in the southwestern Great Artesian Basin, Australia. Austral J Earth Sci 63(7):857–872

Jeanpert J et al (2016) Morphology and distribution of dolines on ultramafic rocks from airborne LiDAR data: the case of southern Grande Terre in New Caledonia (SW Pacific). Earth Surf Proc Landf 41(13):1854–1868

Jennings JN (1975) Doline morphometry as a morphogenetic tool: New Zealand examples. NZ Geogr 31(1):6–28

Jennings JN (1985) Karst geomorphology (2nd edn). Blackwell, Oxford, p 304

Jennings JN, Sweeting MM (1963) The limestone ranges of the Fitzroy Basin, Western Australia. Bonner Geogr Abhand 32:60

Jensen JH (1987) Valley poljes in Florida karst. In: Beck BF, Wilson WL (eds) Karst hydrogeology: engineering and environmental implications. Balkema, Rotterdam, pp 47–51

Keskin I, Yılmaz I (2016) Morphometric and geological features of karstic depressions in gypsum (Sivas, Turkey). Environ Earth Sci 75:1–14

Klimchouk A, Forti P, Cooper A (1996) Gypsum karst of the world: a brief overview. Int J Speleol 25(3–4):159–181

Knez M, Slabe T (2011) The shape and rock relief of pillars in the Naigu Stone Forest. In: Knez M, Liu H, Slabe T (eds), South China Karst II. ZRC, Ljubljana, pp 9–17

Kusumayudha SB, Setiawan J, Ciptahening AN, Septianta PD (2015) Geomorphologic model of Gunungsewu karst, Gunung Kidul Regency, Yogyakarta Special Territory, Indonesia: the role of lithologic variation and geologic structure. J Geol Resour Eng 3(1):1–7

Kuzucuoğlu C (2019) Geomorphological landscapes in the Konya plain and surroundings. In: Kuzucuoğlu C, Çiner A, Kazanci N (eds) Landscapes and landforms of Turkey. Springer, Cham, pp 353–368

Lamb MP, Howard AD, Johnson J, Whipple KX, Dietrich WE, Perron JT (2006) Can springs cut canyons into rock? J Geophys Res: Planets 111:E07002

Lamplugh GW (1902) Calcrete. Geol Mag 9(12):575

Land AL, Johnson KS, Neal JT (2003) Evaporite karst and regional groundwater circulation in the Lower Pecos Valley of Southeastern New México. Evaporite karst and engineering/environmental problems in the United States. Oklahoma Geol Surv Circ 109:227–232

Lang J, Lucas G (1970) Contribution a l'étude de biohermes continentaux; barrages des lacs de Band-e-Amir, Afghanistan Central. Bull Soc Geol France 7(5):834–842

Lipar M, Ferk M (2015) Karst pocket valleys and their implications on Pliocene-Quaternary hydrology and climate: Examples from the Nullarbor Plain, southern Australia. Earth Sci Rev 150:1–13

Lu G, Zheng C, Donahoe RJ, Lyons WB (2000) Controlling processes in a $CaCO_3$ precipitating stream in Huanglong Natural Scenic District, Sichuan, China. J Hydrol 230(1–2):34–54

Luo W et al. (1997) Ground-water sapping processes, western desert, Egypt. Geol Soc Am Bull 109(1):43–62

Lv C et al. (2023) Insights into Alpine-Karst-Type tufa deposits in geological environmental records: a case study of the calcareous tufa profile of the Jiuzhaigou Natural Reserve on the eastern margin of the Tibetan Plateau. Minerals 13(1):120

Lyew-Ayee P (2010) The Cockpit Country of Jamaica: an island within an island. In: Migon P (ed) Geomorphological landscapes of the World. Springer, Dordrecht, pp 69–77

Lyew-Ayee P, Viles HA, Tucker GE (2007) The use of GIS-based digital morphometric techniques in the study of cockpit karst. Earth Surface Proc Landf 32(2):165–179

Lyew-Ayee P (2004) Digital topographic analysis of cockpit karst: a morpho-geological study of the Cockpit Country region, Jamaica (Doctoral dissertation, University of Oxford)

Maizels JK (1987) Plio-Pleistocene raised channel systems of the western Sharqiya (Wahiba), Oman. Geol Soc Lond Spec Publ 35(1):31–50

McLaren SJ (2007) Aeolianite. In: Nash DJ, McLaren SJ (eds) Geochemical sediments and landscapes. Blackwell, Oxford, pp 144–172

Migoń P, Gutiérrez F, Parenti C, Sevil J (2023) Ebro Valley gypsum escarpment near Zaragoza (NE Spain)—combination of highly valuable rock record, dynamic geomorphosites and associated cultural heritage. Geoheritage 15(4):110

Miller RP (1937) Drainage lines in bas-relief. J Geol 45(4):432–438

Mitchell CW, Willimott SG (1974) Dayas of the Moroccan Sahara and other arid regions. Geogr J 140:441–453

Moreno-Pérez PA, Hernández-Téllez M, Bautista-Gálvez A (2021) In danger one of the largest aquifers in the world, the great Mayan aquifer, based on monitoring the cenotes of the Yucatan Peninsula. Arch Environ Contam Toxicol 81:189–198

Mottershead D (1996) A study of solution flutes (rillenkarren) at Lluc, Mallorca. Z Geomorph Suppl 103:215–241

Mousa FA, Ruban DA, El-Hassan MMA, Sallam ES (2024) Late Mesozoic–Cenozoic geoheritage resources of the Kharga Oasis (Egypt): novel assessment, exploitation perspectives, and policy implications. Resources Policy 91:104959

Mylroie JE, Carew JL, Moore AI (1995) Blue holes: definition and genesis. Carbonates Evaporites 10:225–233

Nelson ST, Karlsson HR, Paces JB, Tingey DG, Ward S, Peters MT (2001) Paleohydrologic record of spring deposits in and around Pleistocene pluvial Lake Tecopa, southeastern California. Geol Soc Am Bull 113:659–670

Osborne LAR, Weliange WS, Jayasingha P, Dandeniya AS, Algiriya APP, Pogson RE (2013) Caves and karst-like features in Proterozoic gneiss and Cambrian granite, southern and central Sri Lanka: an introduction. Acta Carsologica 42(1):25–48

Parry B (2020) The provenance of the Norber erratics, North Yorkshire, UK. Proc Yorks Geol Soc 63(2):63–75

Paterson K (1977) Scarp-face dry valleys near Wantage, Oxfordshire. Trans Inst Brit Geogr 2:192–204

Pentecost A (2005) Travertines. Springer, Heidelberg, p 445

Playford PE (1980) Devonian "Great Barrier Reef" of Canning Basin, Western Australia. AAPG Bull 64(6):814–840

Playford PE, Hocking RM, Cockbain AE (2009) Devonian reef complexes of the Canning Basin, Western Australia. Geol Surv W Austral Bull 145:444

Radišić M, Rubinić J, Ružić I, Brozinčević A (2021) Hydrological system of the Plitvice Lakes—Trends and changes in water levels, inflows, and losses. Hydrology 8(4). https://doi.org/10.3390/hydrology8040174

Reeves CC (1970) Origin, classification, and geologic history of caliche on the southern High Plains, Texas and eastern New Mexico. J Geol 78(3):352–362

Reeves CC (1971) Relations of caliche to small natural depressions, southern High Plains, Texas and New Mexico. Geol Soc Am Bull 82(7):1983–1988

Reeves CC (1976) Caliche: origin, classification, morphology and uses. Estacado Books, Lubbock, p 233

Reeves CC (1983) Pliocene channel calcrete and suspenparallel drainage in West Texas and New Mexico. In: Wilson RCL (ed) Residual deposits: surface related weathering processes and materials. Blackwell Scientific, Oxford, pp 179–183

Reynard E, Morand S, Ammann T (2003) Protection et mise en values touristique d'un site géomorphologique: la région du Sanetsch (Valais, Suisse). Lausanne Inst Geogr Trav Recher 24:35–52

Rieger T (1992) Calcareous tufa formations. Searles Lake and Mono Lake. California Geol 45(4):99–109

Ritter B, Albert R, Rakipov A, Van der Wateren FM, Dunai TJ, Gerdes A (2023) Late Neogene terrestrial climate reconstruction of the central Namib Desert derived by the combination of U-Pb silcrete and terrestrial cosmogenic nuclide exposure dating. Geochronology 5(2):433–450

Roberts CR, Mitchell CW (1987) Spring mounds in southern Tunisia. Geol Soc Lond Spec Pub 35(1):321–334

Salomon JN (2011) A mysterious karst: the "Chocolate Hills" of Bohol (Philippines). Acta Carsologica 40(3):424–429

Salomon JN (2009) The tsingy karren fields of Madagascar. In: Ginés A, Knez M, Slabe T, Dreybrodt W (eds) Karst rock features. Karren sculpturing. Založba ZRC, Ljubljana, pp 411–422

Schellmann G, Radke U (2004) The marine Quaternary of Barbados. Kölner Geogr Arb 81:137

Schneider G (2008) The roadside geology of Namibia. Gebrüder Borntraeger, Berlin and Stuttgart, p 294

Scholl DW (1960) Pleistocene algal pinnacles at Searles Lake, California. J Sed Res 30(3):414–431

Searle M (2019a) Sharkiyah, Eastern Oman Mountains. In: Searle M, Geology of the Oman Mountains, Eastern Arabia. Springer, Cham, pp 349–378

Searle M (2019b) Dhofar and the Frankincense Coast. In: Searle M, Geology of the Oman Mountains, Eastern Arabia. Springer, Cham, pp 411–428

Sharp M, Gemmell JC, Tison JL (1989) Structure and stability of the former subglacial drainage system of the Glacier de Tsanfleuron, Switzerland. Earth Surf Proc Landf 14(2):119–134

Smith DI, Drew DP (eds) (1975) Limestones and caves of the Mendip Hills. David and Charles, Newton Abbot, p 424

Smith JR, Giegengack R, Schwarcz HP (2004a) Constraints on Pleistocene pluvial climates through stable-isotope analysis of fossil-spring tufas and associated gastropods, Kharga Oasis, Egypt. Palaeogeogr Palaeoclim Palaeoecol 206(1–2):157–175

Smith JR et al. (2004b) A reconstruction of Quaternary pluvial environments and human occupations using stratigraphy and geochronology of fossil-spring tufas, Kharga Oasis, Egypt. Geoarchaeology 19(5):407–439

Soldati M (2010) Dolomites: the spectacular landscape of the 'Pale Mountains.' In: Migoń P (ed) Geomorphological landscapes of the world. Springer, Dordrecht, pp 191–199

Song L, Liang F (2009) Two important evolution models of Lunanshilin karst. In: Ginés A, Knez M, Slabe T, Dreybrodt W (eds) Karst rock features. Karren sculpturing. Založba ZRC, Ljubljana, pp 453–459

Sperling CHB, Goudie AS (1975) The miliolite of western India: a discussion of the aeolian and marine hypotheses. Sed Geol 13(1):71–75

Sperling CHB, Goudie AS, Stoddart DR, Poole GG (1977) Dolines of the Dorset chalklands and other areas in southern Britain. Trans Inst Brit Geogr 2(2):205–223

Steinemann O et al. (2020) Quantifying glacial erosion on a limestone bed and the relevance for landscape development in the Alps. Earth Surf Proc Landf 45(6):1401–1417

Stepišnik U, Ilc Klun M, Repe B (2017) Assessment of educational potential of geodiversity on example of Cerknica Polje, Slovenia. Dela 47(1):5–39

Stoddart DR, Cann JR 1(965) Nature and origin of beach rock. Journal of Sedimentary Research 35(1):243–247

Stoddart DR, Spencer T, Scoffin TP (1985) Reef growth and karst erosion on Mangaia, Cook Islands: a reinterpretation. Z Geomorph Suppl 57:121–140

Sweeting MM (1972) Karst landforms. Macmillan, London, p 362

Sweeting MM (1995) Karst in China: its geomorphology and environment. Springer, Heidelberg, p 277

Thiry M, Ben Brahim M (1997) Silicifications de nappe dans les formations carbonatées tertiaires du piedmont atlasique (Hamada du Guir, Maroc). Geodinam Acta 10(1):12–29

Thomas TM (1963) Solution subsidence in south-east Carmarthenshire and south-west Breconshire. Trans Inst Brit Geogr 33:45–60

Tihansky AB (1999) Sinkholes, west-central Florida. Land subsidence in the Uited States. US Geol Surv Circ 1182:121–140

Trudgill ST (1979) Surface lowering and landform evolution on Aldabra. Phil Trans Roy Soc Lond, B 286:35–45

Uagod R, Avelar A, Netto ALC (2011) Karstic morphology control in non-carbonate rocks: Santana basin, middle Paraiba do Sul river valley. Brazil. Z Geomorph 55(1):1–13

Upchurch S, Thomas MS, Alfieri MC, Fratesi B, Dobecki TL (2019) The karst systems of Florida: understanding karst in a geologically young terrain. Springer, Dordrecht, p 1202

Veress M (2020) Karst types and their karstification. J Earth Sci 31(3):621–634

Veress M, Toth G, Zentai Z, Schläffer R (2009) The Ankarana tsingy and its development, Carpathian. J Earth Environ Sci 4(1):95–108

Veress M, Telbisz T, Tóth G, Lóczy D, Ruban DA, Gutak JM (eds) (2019) Glaciokarsts. Springer, Cham, p 516

Viles HA, Goudie AS (1990) Reconnaissance studies of the tufa deposits of the Napier Range, NW Australia. Earth Surf Proc Landf 15(5):425–443

Viles HA, Pentecost A (2007) Tufa and travertine. In: Nash DJ, McLaren SJ (eds) Geochemical sediments and landscapes. Blackwell, Oxford, pp 173–199

Viles HA, Taylor MP, Nicoll K, Neumann S (2007) Facies evidence of hydroclimatic regime shifts in tufa depositional sequences from the arid Naukluft Mountains, Namibia. Sed Geol 195(1–2):39–53

Waltham T (ed) (1974) The limestones and caves of north-west England. David and Charles, Newton Abbott, p 477

Waltham T (2005) Karst and caves of Ha Long bay. Speleogenesis Evol Karst Aquifers 3(2):1–9

Waltham T (2008) Fengcong, fenglin, cone karst and tower karst. Cave Karst Sci 35(3):77–88

Waltham T (2010) Guangxi karst: the fenglin and fengcong karst of Guilin and Yangshuo. In: Migoń P (ed) Geomorphological landscapes of the World. Springer, Dordrecht, pp 293–302

Waltham T (2015) Large collapse sinkholes, old and new, in the Obruk Plateau, Turkey. Cave Karst Sci 42(3):125–130

Waltham T (2019) Karst and caves of Palau. Cave Karst Sci 36(3):76–80

Waltham AC, Simms M, Farrant AR, Goldie HS (eds) (1997) Karst and caves of Great Britain. Chapman and Hall, London, p 358

Waltham T, Bell F, Culshaw M (2005) Sinkholes and subsidence: Karst and cavernous rocks in engineering and construction. Springer, Berlin, p 384

Watson RA et al (2019) Sinkholes and uvalas in evaporite karst: spatio-temporal development with links to base-level fall on the eastern shore of the Dead Sea. Solid Earth 10(4):1451–1468

Whiteman CA, Haggart BA (2018) Chalk landforms of Southern England and Quaternary landscape development. Proc Geol Ass:doi. https://doi.org/10.1016/j.pgeola.2018.05.002

Willems L et al (2008) Karsts in sandstones and quartzites of Minas Gerais, Brazil. Cadernos do Laboratorio Xeolóxico de Laxe 33:127–138

Williams PW (2004) Karst. In: Goudie AS (ed) Encyclopedia of geomorphology. Routledge, London, pp 586–594

Wilson P, Barrows TT, Lord TC, Vincent PJ (2012) Surface lowering of limestone pavement as determined by cosmogenic (36Cl) analysis. Earth Surf Proc Landf 37(14):1518–1526

Wirthmann A (2000) Geomorphology of the tropics. Springer, Heidelberg, p 314

Woodhead JD et al (2019) The antiquity of Nullarbor speleothems and implications for karst palaeoclimate archives. Sci Reports 9(1):603

Wright JS (2000) Tufa accumulations in ephemeral streams: observations from the Kimberley, north-west Australia. Austral Geogr 31(3):333–347

Wright VP (2009) Calcrete. In: Nash DJ, McLaren SJ (eds) Geochemical sediments and landscapes. Blackwell, Oxford, pp 10–45

Xuewen Z, Weihai C (2006) Tiankengs in the karst of China. Speleogen Evol Karst Aquifers 4(1):1–18

Yizhaq H, Ish-Shalom C, Raz E, Ashkenazy Y (2017) Scale-free distribution of Dead Sea sinkholes: observations and modeling. Geophys Res Lett 44(10):4944–4952

Zerboni A et al (2020) Geomorphology of the Jebel Qara and coastal plain of Salalah (Dhofar, southern Sultanate of Oman). J Maps 16(2):187–198

Index

A. Goudie, *Rocky Landscapes of the World with Google Earth*, https://doi.org/10.1007/978-3-031-77418-8

GPSR Compliance

The European Union's (EU) General Product Safety Regulation (GPSR) is a set of rules that requires consumer products to be safe and our obligations to ensure this.

If you have any concerns about our products, you can contact us on ProductSafety@springernature.com

In case Publisher is established outside the EU, the EU authorized representative is:

Springer Nature Customer Service Center GmbH
Europaplatz 3
69115 Heidelberg, Germany

Batch number: 10731528

Printed by Printforce, the Netherlands